Cambridge Elements ≡

Elements of Aerospace Engineering
edited by
Vigor Yang
Georgia Institute of Technology
Wei Shyy
Hong Kong University of Science and Technology

DISTINCT AERODYNAMICS OF INSECT-SCALE FLIGHT

Csaba Hefler
Hong Kong University of Science and Technology

Chang-kwon Kang
University of Alabama in Huntsville

Huihe Qiu
Hong Kong University of Science and Technology

Wei Shyy
Hong Kong University of Science and Technology

CAMBRIDGE
UNIVERSITY PRESS

CAMBRIDGE
UNIVERSITY PRESS

University Printing House, Cambridge CB2 8BS, United Kingdom

One Liberty Plaza, 20th Floor, New York, NY 10006, USA

477 Williamstown Road, Port Melbourne, VIC 3207, Australia

314–321, 3rd Floor, Plot 3, Splendor Forum, Jasola District Centre, New Delhi – 110025, India

79 Anson Road, #06–04/06, Singapore 079906

Cambridge University Press is part of the University of Cambridge.

It furthers the University's mission by disseminating knowledge in the pursuit of education, learning, and research at the highest international levels of excellence.

www.cambridge.org
Information on this title: www.cambridge.org/9781108812719
DOI: 10.1017/9781108874229

© Csaba Hefler, Chang-kwon Kang, Huihe Qiu and Wei Shyy 2021

First published 2021

A catalogue record for this publication is available from the British Library.

ISBN 978-1-108-81271-9 Paperback
ISSN 2631-7850 (online)
ISSN 2631-7842 (print)

Additional resources for this publication at www.cambridge.org/hefler

Distinct Aerodynamics of Insect-Scale Flight

Elements of Aerospace Engineering

DOI: 10.1017/9781108874229
First published online: April 2021

Csaba Hefler
Hong Kong University of Science and Technology

Chang-kwon Kang
University of Alabama in Huntsville

Huihe Qiu
Hong Kong University of Science and Technology

Wei Shyy
Hong Kong University of Science and Technology

Author for correspondence: Huihe Qiu, meqiu@ust.hk

Abstract: Insect-scale flapping wing flight vehicles can conduct environmental monitoring, disaster assessment, mapping, positioning and security in complex and challenging surroundings. To develop bio-inspired flight vehicles, systematic probing based on the particular category of flight vehicles is needed. This Element addresses the aerodynamics, aeroelasticity, geometry, stability and dynamics of flexible flapping wings in the insect flight regime. The authors highlight distinct features and issues, contrast aerodynamic stability between rigid and flexible wings, present the implications of the wing-aspect ratio, and use canonical models and dragonflies to elucidate scientific insight as well as technical capabilities of bio-inspired design.

Keywords: low-Reynolds-number flight, insect-scale flight, wing-wing interaction, flapping flight stability, flexible flapping wing flight

ISBNs: 9781108812719 (PB), 9781108874229 (OC)
ISSNs: 2631-7850 (online), 2631-7842 (print)

Contents

Publisher's note: The e-book edition of this title contains colour. The colour images are available in pdf format as an online resource for print readers and users of e-reader devices and applications that cannot display colour: www.cambridge.org/hefler

1 Introduction

The goal of the present work is to give an overview of some of the technicalities of insect-scale flight both in nature and in the realm of engineered flyers known as micro air vehicles (MAVs). The topic is undoubtedly vast and complex, to say the least. Keeping in mind the format of the Cambridge Elements series and the interests of our readers, we decided to investigate four topics in detail: aeroelasticity in flapping flight, stability and dynamics of flexible flapping wings, aerodynamic interactions of tandem winged systems based on dragonflies, and the implications of the geometry of flapping wing aerodynamics. First, in this section, we offer some general background related to flapping wing flight.

1.1 Evolution and Variations of Flapping Flight in Nature

The history of flight started 400 million years ago with insects as the first species to develop wings to fly, at about the same time the first plant species extended their branches further towards the sky to form trees and forests (Misof et al. 2014). Due to the lack of informative fossil records, it is unclear how insect wings evolved. Two alternative theories suggest that the wings evolved either from gills of juvenile stages of the insect (Gegenbaur et al. 1878) or from paranotal lobes (small plates on the side of the animal that is part of its exoskeleton) (Forbes 1943; Müller 1877). The next species capable of powered flight were the pterosaurs, appearing 160 million years after the insects. By their sheer size, it is debatable whether pterosaurs could take off from a stationary position. Some speculated that pterosaurs needed to launch their flight from a place of altitude such as a hilltop. They were apparently clumsy during landing as well (Habib 2013; Hone et al. 2015; Witton & Habib 2010). The first, most primitive descendants of birds, the *Archaeopteryx*, show features both reptilian (teeth, long separate fingers with claws, bony whip-like tail) and birdlike (feathers) and lived 160 million years ago (80 million years after the first pterosaurs (Alexander 2015; Norberg 1990). The origin of flight in birds is a much-debated topic (Alexander 2015). The fourth and the last group of animal species to develop powered flight is bats. The earliest fossil records of bats are from the Eocene (about 50 million years ago), from a species already fully capable of flying that shows little distinction from modern bats (e.g. *Icaronycteris* (Novacek 1985)). There is an almost uniform agreement amongst scientists that bats evolved from arboreal, gliding ancestors (Gunnell & Simmons 2012; Norberg 2002), as first proposed by Darwin (1859).

The development of flight was made possible by specific physical adaptations, and the consequent evolutionary refinement has resulted in the rich diversity of flying species we know today. All flying species use the flapping of their wings to generate thrust and lift. In accordance with the physiological constraints of each species, primary flight behaviour and the scaling laws, a variety of flapping patterns (wing stroke trajectories) are observed (Alexander 2002) as illustrated in Figure 1.1. The illustrated stroke patterns are also found in hovering birds and insects adopting nearly horizontal strokes. Hovering birds are observed using the avian stroke (Azuma 2006; Pennycuick 2008) while insects and notably hummingbirds adapt the figure eight or double figure eight pattern (Azuma 2006; Chen et al. 2013).

What we see as locomotion by the flapping of the wings is a complex motion involving flapping (changing of the positional or stroke angle ζ) in the stroke plane; pitching (pronation and supination) – that is the rotation of the wing parallel to the leading edge (LE) described by the angle of attack (AoA) (α) (Figure 1.2) and finally the variation of the deviation angle (χ) (the deviation of the wings from the stroke plane). Note that phi (φ) is used as the flapping positional angle in many studies. However, phi is also reserved as one of the Euler angles describing the body dynamics in standard aircraft dynamics equations of motion as well as in some notable flapping wing MAV dynamics studies (e.g. Orlowski & Girard 2012a, 2012b). Therefore we use zeta (ζ) to represent the flapping angle and uppercase zeta (Z) to represent the amplitude of

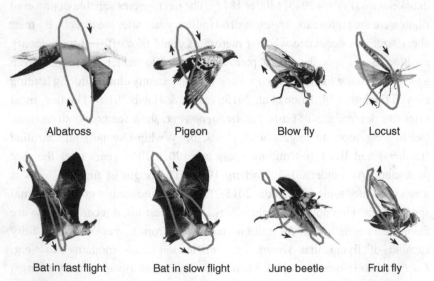

| Albatross | Pigeon | Blow fly | Locust |

| Bat in fast flight | Bat in slow flight | June beetle | Fruit fly |

Figure 1.1 Wingtip trajectories of different flying animals. Some of these flapping patterns, like the ellipsoid avian stroke for medium-sized birds or the typical figure eight stroke for insects and hummingbirds, are also observed in hovering flight with a more horizontally inclined stroke plane. Adapted from Alexander (2002).

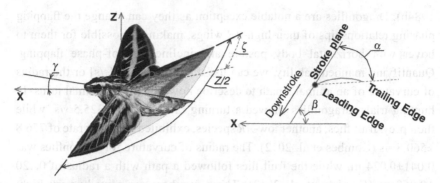

Figure 1.2 The parameters used to describe the flapping kinematics: angle of attack (α); positional or stroke angle (ζ) and the amplitude of the complete flapping cycle (Z); body angle (γ) and the stroke plane angle (β)

flapping angle including both the upstrokes and downstrokes. In addition, natural flyers often set their body angle (γ) to minimize the drag force as well as change the stroke plane angle as an additional control option. The stroke plane is defined by the wing base and the wingtip of the maximum and minimum sweep positions, and the orientation is described by the stroke plane angle (β) (Figure 1.2).

There are more than a million flying insect species, and out of the approximately 13,000 vertebrate species, 9,000 bird species and 1,000 bat species conquered the skies (Shyy et al. 2007). The wings of these animals can take several forms with various specific features. The planform of a wing is most generally described by the aspect ratio (*AR*). The *AR* of the wings of a plane or flying animal is the ratio of its wingspan to its mean chord (or in cases when a single wing is described, the ratio of the length of the wing to its mean chord, being approximately half of the *AR* of an airplane or flying animal using the traditional definition).

High-*AR* wings are typical for high lift generation in gliding and soaring flight of birds (Viscor & Fuster 1987). The *AR* of flapping wings has a major effect on flapping wing performance. Higher-*AR* wings can produce more lift while having a minor effect on thrust (Muniappan et al. 2005). The wandering albatross, *Diomedea exulans*, has high-*AR* (12–15) wings, which makes it the greatest soaring bird; its glide ratio (horizontal velocity/vertical velocity) can reach 21.2 at a cruising airspeed of 16.0 m/s with a sink rate of 0.755 m/s (Pennycuick 2008; Richardson 2011). Low-*AR* wings are often utilized by species cruising and hovering with high agility, like the hawkmoth, *Agrius convolvuli* (*AR* 2.65) (Shyy et al. 2010), the honeybee, *Apis* (*AR* 3.3) (Ellington 1984b), or the hoverfly, *Episyrphus balteatus* (*AR* 4.2) (Ellington

1984b). Dragonflies are a notable exception as they can change the flapping phasing relationships of their high-*AR* wings, making it possible for them to hover with horizontal body posture and inclined out-of-phase flapping. Quantifying manoeuvrability, we can use the turning rate [°/s] or the radius of curvature of an animal's path to describe how quickly an animal turns. In hunting trials, dragonflies showed a turning rate of 914.0±325.5 °/s, while their prey, fruit flies, another low-*AR* species, exhibited a turning rate of 776.8 ±560.8 °/s (Combes et al. 2012). The radius of curvature for dragonflies was 0.041±0.024 m, while the fruit flies followed a path with a radius of 0.120 ±0.068 m (Combes et al. 2012). These numbers are already very high; nevertheless, in some cases where the fruit flies executed evasive manoeuvring, their turning rate exceeded 2000 °/s, allowing them to fly with a curvature as steep as 0.015 m (Combes et al. 2012).

Ancestrally, flying insects had two pairs of wings, as seen in modern dragonflies (Fédrigo & Wray 2010). In some cases, either the forewings or the hindwings evolved to serve an alternate function, and therefore these insects fly with one pair of wings. A well-known example is the *Elytra* of beetles and some bug species which have hardened forewings and protective cover of the hindwings when not in flight (Fédrigo & Wray 2010). Another specialized organ evolved from ancestral hindwings such as flies(*Diptera*), crane flies (*Tipulidae*) or ancestral forewings such as stylops (*Stylops melittae*) are the halteres which help sense body rotation during flight (Dickinson 1999). Closely coupled forewings and hindwings in essence behave as a single pair of wings. For example, hindwing reduction and the mechanical coupling of the wings form a single aerodynamic surface in wasps (*Hymenoptera*), mayflies (*Ephemeroptera*) and several butterfly and moth (*Lepidoptera*) species, while in some cases, the hindwings are completely lost (Dudley 2000). Effectively forming one large aerodynamic surface by both the forewings and hindwings is advantageous for the gliding or sailing flight of migratory species such as the monarch butterfly (*Danaus plexippus*) or the desert locust (*Schistocerca gregaria*).

Conversely, some insects – notably mantises (*Mantodea*), locusts(*Schistocerca*), stoneflies (*Plecoptera*), termites (*Isoptera*), roaches (*Blattaria*) and some dragonfly species – exhibit enlarged hindwing relative to the forewing (Dudley 2000; Wootton 1992). Depending on the flight mission and flapping characteristics, there are relative merits between having functionally one pair or two pairs of wings. Expanded hindwings can help balance the forewings and reduce rotational motion along the longitudinal axis of the wing between upstrokes and downstrokes, which is beneficial for hovering and swift mid-air manoeuvring (Dudley 2000). Insects having functionally two pairs of wings can utilize wing-wake interaction between the forewings and hindwings (see in Section 4) to enhance aerodynamics.

It was also reported that dragonflies are still able, although less swiftly, to fly and hunt with as much as 50% of their hindwing area lost (Combes et al. 2010).

In addition to the given physical dimensions of the wings of different species, active wing morphing is applied by natural fliers within a single flapping cycle. The way birds, bats and insects modify the active area of their wings is fundamentally different (Alexander 2015). Birds deform and twist their wings using specifically evolved bones and musculature that is similar to a human arm; however, birds have more stringent muscle and bone movement during flight. Modifications of the chordwise camber and flexing the wing planform between upstroke and downstroke, twisting, area expansion and contraction and transverse bending are modified by the specialized musculature during flight (Alexander 2015; Shyy et al. 2007; Shyy et al. 2013). Additionally, the feathers and bones of birds are combinations of structural features organized hierarchically from nano- to macroscale that enable a balance between lightweight and bending/torsional stiffness and strength (Sullivan et al. 2017). When birds fly, they extend their wings during downstroke and flex them backwards during upstroke to reduce drag – a flapping behaviour called 'avian stroke' (Azuma 2006; Pennycuick 2008). Bats apply similar strategies to birds when flying; however, bat wings are made of a thin membrane supported by the arm and finger bones and, due to the stretchiness, bat wing membranes deform marginally, reducing the span by about 20% (Hedenström & Johansson 2015; Shyy et al. 2016; Yu & Guan 2015). Wang et al. (2014) modelled numerically the active morphing of wings with a flapping rectangular flat-plate with a dynamically stretching and retracting wingspan (in the flapping cycle, the wingspan varied as a function of time). They found that lift enhancement was achieved by both the effect of changing the wing area and that the effect of the morphing-altered flow structures (leading-edge vortex (LEV) stretching in the upper wing surface, vortex contraction on the bottom wing surface) contributed to the vortex lift modulation. Outer wing separation is also specific to cruising birds and has been studied and found to improve the generated thrust and lift compared to a wing without separation (Mahardika et al. 2011).

Insect wings deform significantly during flight. The observed deformations of insect wings are passive, resulting from aerodynamic and inertial effects. Chordwise flexibility and the associated redistribution of thrust and lift (Kang et al. 2011; Shyy et al. 2010) as well as spanwise flexibility and effective AoA distribution due to shape variation along the wingspan (Kang et al. 2011; Kodali et al. 2017; Shyy et al. 2010) can aid in lift generation and power reductions. These benefits of wing flexibility have received significant interest particularly for hovering flight. Hovering is a costly flight mode as there is no surrounding flow that can enhance lift production (Berman & Wang 2007). Adequate understanding of the interplay between the wing structure and unsteady aerodynamics can decipher the physical mechanism behind insect flight and further contribute to the development of bio-inspired MAVs.

Notably, stability studies in the literature are largely based on the rigid wing framework, leading to the conclusion that the hover equilibrium condition is intrinsically unstable (Orlowski & Girard 2012a; Sun 2014; Taha et al. 2012). Rigid wings possess an unstable oscillatory mode mainly due to their pitch sensitivity to horizontal velocity perturbations. However, if the wing flexibility is included in the analysis, a flapping wing can experience stable hover equilibria (Bluman et al. 2018; Bluman & Kang 2017a). In essence, flexible wings tend to stabilize the unstable mode by passively deforming their wing shape in the presence of perturbations, generating significantly more horizontal velocity damping and pitch rate damping. These results have significant implications and suggest that weight and mechanical properties such as stiffness and geometry influence flight stability of insect-scale flapping wing robots and need to be considered systematically and carefully.

1.2 Governing Equations and Dimensionless Parameters of Flapping Flight

In fluid dynamics, the continuity equation (Eq. 1.1) is used to express the conservation of mass. Additionally, the Navier-Stokes equation (Eq. 1.2), is used to express the conservation of momentum and to describe the motion of the fluid in a continuum:

$$\frac{\partial \rho}{\partial t} + \nabla \cdot (\rho \mathbf{v}) = 0 \tag{1.1}$$

$$\rho \left[\frac{\partial \mathbf{v}}{\partial t} + (\mathbf{v} \cdot \nabla)\mathbf{v} \right] = \nabla \cdot \mathbf{P} + \rho \mathbf{f} \tag{1.2}$$

In these equations, '$\mathbf{v}$' is the velocity vector of the fluid, 't' is the time, 'ρ' is the density of the fluid, '$\mathbf{P}$' is the pressure tensor and '$\mathbf{f}$' is the external body force such as gravity. '∇' denotes the gradient operator.

In the case of flapping wing flight, the fluid can be considered as incompressible Newtonian fluid. In this case, these equations can be simplified as:

$$\nabla \cdot \mathbf{v} = 0 \tag{1.3}$$

$$\rho \left[\frac{\partial \mathbf{v}}{\partial t} + (\mathbf{v} \cdot \nabla)\mathbf{v} \right] = -\nabla p + \mu \nabla^2 \mathbf{v} + \rho \mathbf{f} \tag{1.4}$$

where 'μ' is the dynamic viscosity of the fluid and 'p' is the static pressure.

When considering an experimental set-up to determine quantitative and qualitative features of the flow around flapping wings, another difficulty arises from the fact that flapping flight in nature features small sizes and high-frequency flapping kinematics that make measurement very difficult.

Scaling the experiments to be feasible under conventional laboratory set-ups is an essential tool to the study of flapping flight. Following the concept of dimensional homogeneity, the Navier-Stokes and the continuity equation can be normalized to establish a relationship between the scaled model and the studied system. In our case, we can define a reference velocity 'U', reference pressure 'p_0', reference length 'L', a reference time or frequency '$t = 1/f$' and a reference body force 'g'. The non-dimensional equations (asterisks mark a non-dimensional parameter) appear as:

$$\nabla^* \cdot \mathbf{v}^* = 0 \tag{1.5}$$

$$\left(\frac{fL}{U}\right)\frac{\partial \mathbf{v}^*}{\partial t^*} + (\mathbf{v}^* \cdot \nabla^*)\mathbf{v}^* = -\left(\frac{p_0}{\rho U^2}\right)\nabla^* p^* + \left(\frac{\mu}{\rho UL}\right)\nabla^{*2}\mathbf{v}^* + \left(\frac{gL}{U^2}\right)f^*$$

$$\tag{1.6}.$$

The four dimensionless groups from this equation are: Strouhal number

$$St = \frac{f \cdot L}{U} \tag{1.7},$$

Euler number

$$Eu = \frac{p_0}{\rho U^2} \tag{1.8},$$

Reynolds number

$$Re = \frac{\rho UL}{\mu} \tag{1.9},$$

Froude number

$$Fr = \frac{U}{\sqrt{gL}} \tag{1.10},$$

In the case of flapping flight, the only body mass in the fluid is gravity that is superimposed on the hydrostatic pressure in the flow field. Accordingly, we can define a new pressure $p' = p - \rho gz$, where 'z' is the fluid column on the wing. This simplifies the dimensionless equation, as the Froude number will not be included. Under the conditions of flapping flight, when cavitation is not present, the Euler number has no importance. In this case, 'p_0' can be taken as 'ρU^2' and the Euler number will not appear in the equations (Munson et al. 2012). After these assumptions, the equations appear in the form of:

$$\nabla^* \cdot \mathbf{v}^* = 0$$

$$\left(\frac{\omega L}{U}\right)\frac{\partial \mathbf{v}^*}{\partial t^*} + (\mathbf{v}^* \cdot \nabla^*)\mathbf{v}^* = -\nabla^* p^* + \left(\frac{\mu}{\rho UL}\right)\nabla^{*2}\mathbf{v}^* \qquad (1.11).$$

The Strouhal number is a dimensionless parameter used to describe systems with oscillating flow. In the case of flapping wings, it is defined by the flapping frequency 'f', the flapping amplitude as reference length (that is the root-to-tip wing length 'l' multiplied by the stroke amplitude in radians 'Z') and the forward flight speed 'U_∞' as reference velocity – namely:

$$St = \frac{f \cdot l \cdot Z}{U_\infty} \qquad (1.12).$$

It is the ratio of the flapping speed of the wing versus the speed of the forward flight. It represents the ratio of the flapping speed to the forward flight speed. It gives a measure of the ratio of inertial forces due to the unsteadiness of the flow (local acceleration) to the inertial forces due to changes in velocity from point to point in the flow field (convective acceleration) (Munson et al. 2012). The shedding behaviour of vortices and the vortex dynamics in the wake can also be characterized by the Strouhal number (Shyy et al. 2007; Triantafyllou et al. 2000; Wang 2000), and it offers a measure of propulsive performance (Shyy et al. 2010; Triantafyllou et al. 2000). Natural flyers and swimming species operate in a narrow range of *St* between 0.2 and 0.4, resulting in high propulsive efficiency (Taylor et al. 2003; Triantafyllou et al. 2000). In this *St* range, it was observed that flapping wings and airfoils leave a characteristic path of wake vortices called the reverse von Karman vortex street (Shyy et al. 2013).

In practice, another dimensionless parameter, the *reduced frequency* 'k', is used to describe the unsteadiness in flapping flight systems. The reduced frequency compares the spatial wavelength of the flow disturbance with the chord of the wing. It also gives the ratio between the fluid convection timescale c_m/U and the wing motion timescale $1/f$. The reduced frequency is calculated from the angular speed of the flapping wing ($2\pi f$) multiplied by the reference length that is the mean chord length of the wing (c_m) and divided by the reference velocity, which can be either the wingtip velocity (Eq. 1.13) or the forward flight velocity (Eq. 1.14):

$$k = \frac{\pi f c_m}{U_{ref}} = \frac{\pi f c_m}{U_{tip}} = \frac{\pi}{Z \cdot AR} \tag{1.13}$$

$$k = \frac{\pi f c_m}{U_{ref}} = \frac{\pi f c_m}{U_\infty} = \frac{2\pi St}{Z \cdot AR} \tag{1.14},$$

where AR is calculated using the wing length ($AR = l^2/A$, where 'A' is the wing planform area) and 'Z' is the previously defined flapping amplitude. If using the tip velocity as the reference velocity (Eq. 1.13), the reduced frequency is inversely proportional to the flapping amplitude and the AR of the wing. It is not related to the actual flapping frequency of the wing. Equation 1.14 gives the relationship between St and k that is $AR \cdot Z/2\pi$.

In the case of flapping flight, the mean chord is used as the reference length, and either the freestream velocity or the wingtip velocity is used as the reference velocity. The chord of the wings of natural fliers varies along the length of their wing, in which case the mean chord is determined by dividing the wing planform area with the length of the wing. In the case of animals with two pairs of wings, either the forewing or the hindwing is used to determine the reference length and velocity.

Accordingly, the *Reynolds number* is:

$$Re = \frac{\rho U_{ref} c_m}{\mu} \tag{1.15}.$$

Physically, the Reynolds number gives the ratio between the inertial and viscous forces. Regarding insects, Re ranges from about 10 to 1,000, while in the case of birds, it is about 1,000 to 15,000 (Park & Yoon 2008; Shyy et al. 2016; Shyy et al. 2007). The flapping frequency of a species normally increases as the size decreases (Azuma 2006; Shyy et al. 2013). Relationships of the Reynolds number, typical size, mass and flapping frequency (where applicable) of natural fliers and aircraft are shown in Figure 1.3. The Reynolds number and the reduced frequency directly appear in the Navier-Stokes equation, where 'c_m', 'U' and '$1/f$' are used as the references:

$$\frac{k}{\pi} \frac{\partial \mathbf{v}^*}{\partial t^*} + (\mathbf{v}^* \cdot \nabla^*) \mathbf{v}^* = -\nabla^* p^* + \left(\frac{1}{Re}\right) \nabla^{*2} \mathbf{v}^* \tag{1.16}.$$

Keeping all the dimensionless parameters unchanged when scaling an experiment has some practical limitations, but in most cases, it is possible to scale up

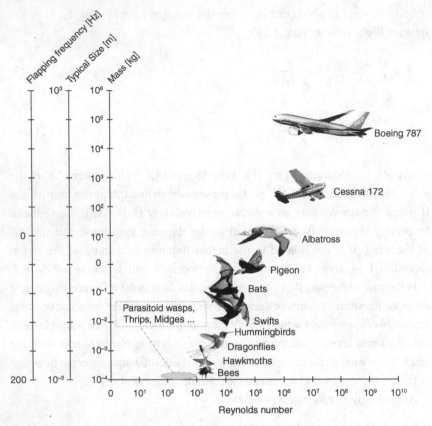

Figure 1.3 The relationships between the Reynolds number and the mass, the size as well as the flapping frequency

a flapping wing system that provides a higher resolution for detailed flow measurements (Cheng et al. 2014; Chowdhury et al. 2019; Maybury & Lehmann 2004; Park & Choi 2012; Zheng et al. 2016a). In practice, not only the physical size of the wing is enlarged, but the flapping frequency is reduced in such set-ups (Figure 1.4 shows three such set-ups). This allows the precise observation of flow dynamics in a time-resolved manner as well as direct force measurement using conventional strain gauge force/torque sensors. In a scaled-up set-up, the fluid density is increased, reducing the ratio of the inertial force in relation to the fluid dynamic forces. Additional to these parameters, the boundary conditions should be kept similar to those in the modelled system. Active or passive wing morphology is one such condition. Similarity in passive wing morphology can be ensured by providing adequate structural flexible properties to the model wing.

A flexible wing can be modelled as a plate that twists and bends in the spanwise and chordwise directions (Shyy et al. 2010; Shyy et al. 2013). The

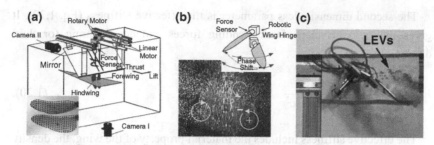

Figure 1.4 Scaled-up experimental set-ups for flow visualization and measurement of flapping wings in water. (a) Particle image velocimetry flow measurement and shape measurement with the direct linear transformation technique, reprinted from Zheng et al. (2016a) with permission from Elsevier. (b) Flow visualization and particle image velocimetry flow measurement by tiny bubbles in oil (Maybury & Lehmann 2004 – reproduced with permission of the *Journal of Experimental Biology*). (c) Dye flow visualization in water (Chowdhury et al. 2019).

transverse displacement of a thin isotropic plate is described by the plate equation (Eq. 1.17):

$$\frac{E t_w^3}{12(1-\nu^2)\rho_f U^2 c^3}\left(\frac{\partial^4 w^*}{\partial x^{*4}}+2\frac{\partial^4 w^*}{\partial x^{*2}y^{*2}}+\frac{\partial^4 w^*}{\partial y^{*4}}\right)-\frac{\rho_w t_w}{\rho_f c}\left(\frac{k}{\pi}\right)^2\frac{\partial^2 w^*}{\partial t^{*2}}=f_e^*$$

(1.17).

'E' is Young's modulus of the wing; 'ν' is Poisson's ratio of the wing; 'ρ_w' and 'ρ_f' represent the density of the wing and the fluid, respectively; 'c' is the wing chord length; 'w' is the transverse displacement of the flexible wing; 't_w' is the wing thickness; 'f_e' is the distributed external force. The plate equation used as the third governing equation for a flexible flapping wing system adds two additional dimensionless parameters to be considered when scaling an experiment:

$$\rho^* = \frac{\rho_w}{\rho_f}$$

(1.18)

$$\Pi_0 = \frac{\rho_w t_w}{\rho_f c}\left(\frac{k}{\pi}\right)^2$$

(1.19).

The *density ratio* (Eq. 1.18) describes the ratio between the equivalent structural density and fluid density; in another form, the *effective inertia* (Eq. 1.19) is the ratio of inertial to aerodynamic forces.

The second dimensionless parameter is the effective stiffness (Eq. 1.20). It gives the ratio between elastic bending forces and aerodynamic (or fluid dynamic) forces:

$$\Pi_1 = \frac{Et^3}{12(1 - \nu^2)\rho_f U^2 c^3} \tag{1.20}.$$

The effective stiffness includes the material property of the wing, the density of the fluid and the geometry of the wing, as well as the velocity as a kinematic parameter; thus it provides a complete set of parameters that affect the deformation of the wing.

Maintaining geometrical similarity is an easy task compared to the difficulty of ensuring that the dimensionless parameters remain unchanged when designing a scaled experiment. The reason is that the scaling laws for rigid and flexible flapping wings alike are often different among the involved dimensionless parameters. Table 1.1 gives a summary of these scaling laws considering both the wingtip velocity and the forward flight velocity as references.

1.3 Unsteady Aerodynamic Mechanisms of Flapping Flight

Within the past century, the enrichment of the theoretical understanding of traditional fixed-wing aerodynamics and the continuous advancements of human aviation technologies have followed each other hand in hand. Aerodynamic characteristics of fixed wings are well understood but are not applicable to flapping flight. The traditional quasi-steady approach cannot explain how flapping insects are capable of staying aloft (Ellington 1984a). There are differences in operation, geometry and structure between the wings of an airplane and the wings of an insect taken as an example. The wing of an airplane generates lift by moving through the fluid (a motion that is generated by the thrust of the engines) at an AoA to the flow direction, deflecting the fluid downwards (Babinsky 2003). Lift is generated by turning a moving fluid, creating pressure gradients across streamlines. The lower surface of the wing pushes the fluid downwards; a force acts on the lower surface of the wing equal to the force needed to change the motion of the fluid flow below the wing. Similarly, the top surface of the wing pulls the fluid flow, which again results in a downward deflection of the flow. There is a pressure difference between the top and bottom surfaces of the airfoil that is on the complete surface of the wing, giving the lift force. The top surface of the airfoil generates typically two-thirds of the lift while the bottom surface accounts for the other third (Babinsky 2003). If the AoA is too large, the flow separates from the upper surface of the wing, resulting in a sudden drop of lift that causes stall, a potentially catastrophic event for an aircraft.

Table 1.1 Dimensionless parameters and their scaling dependency (Shyy et al. 2010)

Dimensionless parameter	Based on flapping wing velocity (U_{tip})		Based on flight speed (U_∞)		
	Length	Frequency	Length	Frequency	Velocity
Reynolds number	c_m^2	F	c_m	Independent	U_∞^1
Reduced frequency	Independent	Independent	c_m	f	U_∞^{-1}
Strouhal number	Independent	Independent	c_m	f	U_∞^{-1}
Effective stiffness	Independent	f^{-2}	Independent	Independent	U_∞^{-2}
Density ratio	c_m^{-2}	Independent	Independent	Independent	Independent

A flapping wing of a natural flyer or an MAV operates differently and under different conditions to rigid wings. (i) In a low-Reynolds-number regime of about 10^1–10^4 for natural flyers versus 10^6–10^7 for commercial aircraft, viscous effects in the flow field become significant. The lift of a fixed wing reduces significantly in this regime due to laminar-turbulent boundary layer instabilities. (ii) Commercial aircraft often have wings of AR 4.5 and larger to maximize efficiency, provided that the structural integrity of the aircraft is not compromised. While high-AR wings provide fuel efficiency by reducing downwash created by the tip vortices (TiVs) (Munson et al. 2012), they also reduce the manoeuvrability of a plane – a trade-off well justifiable for commercial long-distance flights. Manoeuvrability is significantly more important for flying insects than for civilian aircraft. The AR of most insect wings is typically less than 5 (Shyy et al. 2013). In view of this, we may consider an insect wing AR of 1–2 as low (butterflies, moths), 2–4 (honeybee) as intermediate and higher than 4 (dragonflies, damselflies, locusts) as high-AR wings. The lower AR of natural flyers could suggest that the three-dimensional (3D) effects of the flow around flapping wings might not be a handicap but are utilized to gain useful aerodynamic force and manoeuvrability. Structural constraints and evolutionary adaptation other than to maximize aerodynamic functionality might also affect the wing form of natural flyers. (iii) The second geometrical difference between airplane wings and natural wings is their typical cross section. Aircraft wings are streamlined and thick compared to the thin, membrane-like wings of insects or the slim wings of birds and bats. A thin wing with sharp edges easily causes flow separation and sheds vortical structures. Nevertheless, it is found that a thin airfoil with sharp edges can outperform a traditional wing profile at a low Reynolds number (Lentink & Dickinson 2009). (iv) A traditional fixed wing generates lift more steadily as the wings are quasi-steady in the oncoming flow that is generated by the translational motion of the aircraft. The wings of an airplane do not generate the thrust but leverage it. On the other hand, a flapping wing undergoes a high-frequency periodic unsteady motion, rotational motions (flapping and pitching) around the axes of the wing root and a deviation longitudinal to the body of the animal. This complex motion gives place to several unsteady flow phenomena and makes the wing inherently operate under the influence of its own generated wake. (v) Inherently, wings of animals are anisotropic and flexible. Their shape might or might not be actively changed by muscles in the wing, and passively deform according to the forces acting on the wing (Kang et al. 2011; Shyy et al. 2010). These features of flapping wings create a number of unconventional methods for aerodynamic force generation via unsteady flow mechanisms such as delayed stall of the LEV, clap and fling, wake capture, rapid pitch rotation, TiV or passive shape deformation.

The *delayed stall associated with the LEV* is a mechanism that occurs when there is a sufficiently large AoA between the incident flow and the LE of the wing. As the wing stroke progresses, the effective AoA (Figure 1.5) of a flapping wing is the angle between the wing chord line and the direction of the incident flow that is composed of two velocity components, the flight velocity of the animal and the circumferential velocity of the flapping wing. This effective AoA dynamically changes throughout the flapping cycle and along the span of the wing (Figure 1.5). As the wings flap faster, the ratio between the circumferential velocity and the flight velocity increases, resulting in a substantially larger AoA than in the case of a fixed-wing aircraft.

The existence of an LEV is the most important factor that contributes to the success of flapping flight in the low-*Re* regime (Dickinson & Gotz 1993; Ellington et al. 1996; Shyy & Liu 2007). Figure 1.6 shows the streamline patterns at three Reynolds numbers: (a) corresponds to a hawkmoth hovering at *Re* ~ 6,000, (b) corresponds to a fruit fly at *Re* = 120, and (c) corresponds to a thrips at *Re* = 10. At *Re* = 6,000, an intense, conical LEV core is observed on the paired wings with a substantial spanwise flow at the vortex core, breaking down at approximately three-quarters of the span towards the tip. At *Re* = 120, the vortex no longer breaks down and is connected to the TiV. The spanwise flow at the vortex core weakens as the Reynolds number is

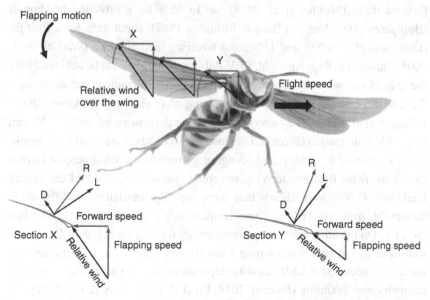

Figure 1.5 Components of incident velocity that define the local angle of attack along the wingspan and the resulting forces. D = drag force, L = lift force, R = resultant force. Adapted for wasp from Alexander (2002).

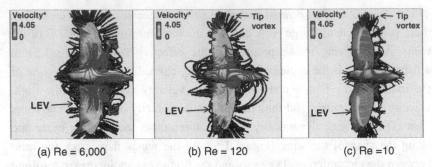

(a) Re = 6,000	(b) Re = 120	(c) Re =10

Figure 1.6 Illustrations of leading-edge vortex structures at different Reynolds numbers. The results are calculated by computational fluid dynamics (Shyy & Liu 2007). From left to right: hawkmoth hovering at Re = 6,000, fruit fly at Re = 120 and thrips at Re = 10.

lowered. Further reducing the Reynolds number to $Re = 10$, a vortex ring connecting the LEV, the TiV and the trailing vortex is observed; the flow structure shows more of a cylindrical than a conical form. Inspecting the momentum equation, one can see that together they are likely responsible for the LEV stability.

It is estimated that the prolonged presence of LEV on the wing can contribute to more than 40% of the total force for a small nectar-feeding bat during slow forward flight (Muijres et al. 2008), up to 65% for a hovering hawkmoth (Bomphrey 2005; Van den Berg & Ellington 1997), about 45% for a fruit fly (Dickinson et al. 1999) and 15% for a hovering hummingbird (Warrick et al. 2005). Inherent in flapping flight, LEV stability is suggested to be achieved with the help of spanwise flow resulting from the pressure gradient and centrifugal force or the flapping motion around the wing pivot (Birch & Dickinson 2001; Ellington et al. 1996). They also suggested that the downwash of the TiVs can help LEV stabilization (Birch & Dickinson 2001). Others reported LEV breakdown in case of a pitching and plunging motion of the wing despite having strong spanwise flow due to wing sweeping (Beem et al. 2012). Lentink and Dickinson (2009) have shown that for a steadily translating fruit-fly wing, downwash from the TiVs or spanwise flow solely is not enough to stabilize the LEV. On the other hand, for a continuously rotating model, the LEV can be stabilized in the case when strong Coriolis and centripetal accelerations are present. It seems that LEV stability depends on the Reynolds number, wing geometry and flexibility (Fu et al. 2014; Fu et al. 2017; Shyy & Liu 2007).

When the wing accelerates and decelerates, the surrounding air moves accordingly, resulting in a force that acts on the wing. This effect is generally referred to as the *added mass effect*, and is often modelled mathematically as a time-variant increase in the inertia of the wing (Lehmann 2004; Sane 2003; Sane & Dickinson 2001). The added mass and other important unsteady effects are summarized and schematically presented in Figure 1.7.

The *clap and fling mechanism* was first reported by Weis-fogh (1973), who studied the flight of the chalcid wasp. During clap and fling, first the wings' LEs touch together at the end of the dorsal stroke of the wasp. Then, the contact from the LE spreads towards the trailing edge (TE) of the wings. This pushes fluid out from the decreasing space between the wings and generates a propulsive jet (Lehmann et al. 2005). Very soon after, the wings start to move away from each other, again starting from the LE. The sudden drop of pressure between the separating wing membranes pulls fluid in the gap, which generates an initial circulation and eventually LEV formation (Lehmann et al. 2005; Miller & Peskin 2005; Weis-fogh 1973). The formed vortices are not accompanied by a pair of starting vortices, thus net lift production is immediately established by the bound vortex pair on the LE of the wings (Weis-fogh 1973). Insects utilizing clap and fling are found to generate, on average, 25% more

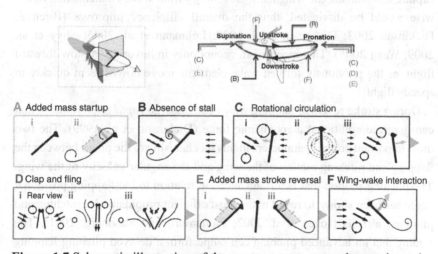

Figure 1.7 Schematic illustration of the most common unsteady aerodynamic mechanisms of insect-scale flapping flight. The relative timings of the mechanisms during each wingbeat are indicated in the upper-right diagram. In all panels, the wing is represented by its chord, drawn from a point of view directed along the spanwise axis of the wing, as shown in the upper-left diagram. Airflow direction is indicated by black arrows, induced velocity is indicated by dark blue arrows, and net forces are represented by light blue arrows. Black triangles indicate the top surface of the leading edge. Chin and Lentink (2016) after Sane (2003) – reproduced with permission of the *Journal of Experimental Biology*.

muscle-mass-specific lift than those that do not (Marden 1987). Miller and Peskin (2005) have shown that lift augmentation is elevated in the lower-Reynolds-number flight regime. Clap and fling is often associated with the smallest flying insects such as parasitoid wasps (*Hymenoptera*) and thrips (*Thysanoptera*) (Ellington 1984c; Miller & Peskin 2005; Santhanakrishnan et al. 2014). Other than these small insects, species that adopt high-amplitude flapping are observed to utilize at least partially the clap and fling mechanism, such as fruit flies (Lehmann et al. 2005), locusts (Cooter & Baker 1977), damselflies (Rudolph 1976a, 1976b; Wakeling & Ellington 1997) and butterflies (Srygley & Thomas 2002).

The flapping wings undergo a periodic motion that facilitates *wake capture*. During stroke reversal and as the consecutive stroke commences, the wing moves through flow induced by the previous stroke. When an LEV is formed by the previous stroke, it sheds normally during stroke reversal as the wing decelerates. As the wing commences its consecutive stroke, the wing collides with the shed LEV of the previous stroke, adding extra lift. During this wake capture mechanism, the wing harnesses energy from wake elements that otherwise would be dissipated, thus the overall efficiency improves (Birch & Dickinson 2003; Dickinson et al. 1999; Lehmann et al. 2005; Shyy et al. 2009; Wang 2005). This mechanism occurs only in hovering or slow forward flight, as the previously formed wake elements move downstream quickly in speedy flight.

During stroke reversal, flapping wings can undergo *rapid pitch rotation* that can generate additional aerodynamic force (Dickinson et al. 1999). The two most important control parameters in this mechanism are the phase between the flapping and pitching motion of the wing and the angular velocity of the wing. The pitching motion that is 90° advanced in relation to the flapping positional angle has been shown to result in the most efficient propulsion for pitching and plunging airfoils (Jones et al. 2002; Ramamurti and Sandberg, 2001). The finding that an advanced pitching can outperform a delayed pitching flapping set-up can be explained with the Magnus effect. If the wing pitches before the stroke reversal, it undergoes a rotation that increases the vorticity around the wing that leads to lift augmentation (Shyy et al. 2010; Shyy & Liu 2007; Sun & Tang 2002). Sane and Dickinson (2002) found that the resulting lift peak is proportional to the angular velocity of the wing.

In a steady flow around a rigid wing aircraft, a vortex pair extends from the tip of the wings. These *tip vortices* are intrinsic to a finite wing. An associated phenomenon is the downwash, which alters the effective AoA of the wing. Compared to the lift on an idealized two-dimensional (2D) wing, the lift on a finite wing is lower and induces additional drag (Anderson 2011). Differently, in flapping wing flight, TiVs can benefit flight efficiency in several ways (Shyy et al. 2010). Tip vortices can create a low-pressure area on the suction side of the tip region of the flapping wings, and an advantageous interaction with the LEV is found in previous studies (Aono et al. 2008; Aono et al. 2009; Ramamurti & Sandberg 2007; Shyy et al. 2009). These effects, however, are found to be sensible for the flapping kinematics and wing geometry. For lower-*AR* wings, the TiV could have a strong downwash effect that reduces lift (Taira & Colonius 2009) or that makes little to no impact depending on the phasing between the flapping and pitching motion of the wing (Aono et al. 2009).

Wings of animals are flexible in nature and can go through substantial *passive shape deformation* (Shyy et al. 1999; Wootton 1981, 1992). Wings deform both in the spanwise and chordwise directions and are prone to twist along their span. The chordwise deformation of the wing essentially means that the TE lags the LE while the wing is flapping. A flexible TE sheds less vorticity in its wake. Chordwise flexibility can help maintain a captured vortex or a shedding LEV on the wing surface (Hefler et al. 2018; Ishihara et al. 2009a). Chordwise flexibility improves thrust generation and reduces drag (Heathcote & Gursul 2007; Heathcote et al. 2004). Insect wings have a strong LE that makes spanwise deformation of the wing smaller than its chordwise deformation. Nevertheless, spanwise deformation can prevent the burst of the LEV at the outer span (Heathcote et al. 2004; Heathcote et al. 2008). Wing twisting results in a gradual decrease of the local AoA towards the wingtip that could cause local effects on lift generation (Roccia et al. 2017). Inertial forces during stroke reversal cause a passive pitching of flexible flapping wings (Bergou et al. 2007; Dickinson et al. 1999). Vanella et al. (2009) presented that in the case when the flapping frequency was less than the wings' natural frequency, the wing undergoes an advanced passive pitching that results in increased lift generation. The effect of wing flexibility is discussed in detail in Section 2.

1.4 Micro Air Vehicles

Micro air vehicles are often characterized as those of sizes smaller than 15 cm (Mcmichael & Francis 1997; Shyy et al. 1999). Early development of MAVs was mainly driven by interest in intelligence gathering and in the monitoring of air quality or hazards (chemical, nuclear or pollution) (Davis et al. 1996; Guizzo 2011; MacRae 2016; Sutherland 2011). Of course, we are now capable of designing and building MAVs of much smaller sizes and more agile perform-ance. Table 1.2 compares a few well-known flapping wing MAVs. Please note that the presented values in Table 1.2 are approximations as some of these MAVs are under current development and improvement.

Table 1.2 Comparison of recently developed flapping wing MAVs (Figures are reproduced with permission and cited in the paragraph following the table.)

	Microbat	R. jellyfish	Robobee	Delfly micro	Robotic h. bird
Weight	12.5 g	2.1 g	0.08 g	3.07 g	20.4 g
Wingspan	229 mm	180 mm	30 mm	100 mm	170 mm
Frequency	30 Hz	20 Hz	120 Hz	30 Hz	34 Hz
Endurance	42 s	–	–	3 min	–
Hover	No	Yes	Yes	No	Yes
Reynolds number	2.8×10^4	3.0×10^2	2.5×10^3	5.2×10^3	1.7×10^4
Strouhal number	0.375	–	–	0.107	0.143

The microbat (Pornsin-Sirirak et al. 2001) is a small ornithopter with wings fabricated using microelectromechanical (MEMS) technology that allows for precise repeatability, size control, weight minimization, mass production and fast turnaround time. The robotic jellyfish (Ristroph & Childress 2014) uses four flapping winglets in an arrangement that mimics the propulsion of jellyfish rather than that of insects. Without relying on additional aerodynamic surfaces or feedback control, the robotic jellyfish achieves a self-righting flight (Ristroph & Childress 2014). Robobee (Wood 2007) utilizes piezo-electric actuation of the wings, which allows for substantial weight reduction and high-frequency flapping. Newer prototypes of "Robobees" are now able to fly untethered using a solar-powered four-winged set-up (Jafferis et al. 2019) as well as swim underwater (Chen et al. 2017). Delfly micro (Deng et al. 2015; Lentink et al. 2010) and Delfly nimble (Karásek et al. 2018) are amongst the smallest ornithopters today that are capable of highly acrobatic flight due to yaw torque coupling. Delfly also utilizes the clap and fling mechanism for efficient force generation. The robotic hummingbird developed at Purdue University is capable of hovering and attitude stabilization in free flight using at-scale fully uncoupled wings (Tu et al. 2020a; Tu et al. 2020b). These examples highlight well the versatility of MAV designs and applications. Further discussions on some of the future prospects regarding the applications of MAVs are in Section 6.

2 Flexible Wing Aeroelasticity

The wings of flying animals are flexible in their structures (Alben et al. 2002; Shyy et al. 1999). The aeroelastic characteristics of a wing are a main interest in biological and bio-inspired flight. It has been, for example, shown that wing deformations (Walker et al. 2009; Young et al. 2009) can enhance the force generation and efficiency of a locust operating at $Re \approx 4 \times 10^3$ (Young et al. 2009) or a hawkmoth operating at $Re \approx 6.3 \times 10^3$ (Nakata & Liu 2012). Moreover, several studies have suggested that insect wing rotations may be passive (Ennos 1988; Ishihara et al. 2009b), meaning that the resulting rotation is due to a dynamic balance between the wing inertial force, elastic restoring force and fluid dynamic force.

It is established that shape adaptation associated with flexibility can affect the effective AoA as well as wing shape and hence the aerodynamic outcome (Shyy et al. 2010; Shyy et al. 1999). Chordwise flexibility can substantially adjust the projected area normal to the flight trajectory or wing motion via shape deformation, hence redistributing thrust and lift (Ramananarivo et al. 2011; Shyy et al. 2010). Katz and Weihs (1978) conducted a parametric study of large-amplitude oscillatory propulsion, with special emphasis on the effect of chordwise flexibility. They reported that flexibility can increase the propulsive efficiency by up to 2% while causing small decreases in the overall thrust, compared with similar motion with rigid foils. Spanwise flexibility in the forward flight creates shape deformation from the wing root to the wingtip, resulting in varied phase shift and effective AoA distribution along the wingspan (Gordnier et al. 2013; Kodali et al. 2017). In addition, the increased acceleration of the wing due to the flexibility enhances force generation (Alben & Shelley 2005; Dewey et al. 2013; Eldredge et al. 2010; Gordnier & Attar 2014; Kang & Shyy 2013; Michelin & Smith 2009; Ramananarivo et al. 2011; Shelley & Zhang 2011; Shyy et al. 2013; Spagnolie et al. 2010; Vanella et al. 2009; Vargas et al. 2008; Wu et al. 2011; Yin & Luo 2010) as well as survivability in the event of a collision (Alben et al. 2002; Bushnell & Moore 1991; Mountcastle & Combes 2014; Shelley & Zhang 2011). However, our understanding of fluid physics and the resulting structural dynamics is to be further developed in order to explain all the salient features of this coupled fluid-structure system including the relationship between structural flexibility, resulting flapping kinematics and unsteady aerodynamics.

2.1 Non-dimensional Parameters and Scaling to Characterize Flexible Wings

Insect wings are formed by a cuticle membrane of varied thickness within a complex network of tubular supporting veins, resulting in non-homologous structures of considerable strength and high flexibility. These wings can endure collisions and tearing without compromising the overall integrity (Combes 2010; Parle et al. 2017; Schmidt et al. 2020; Wootton 1992). The main longitudinal veins are not only structural elements but transferring fluid, oxygen, and sensory information (Pass 2018). The secondary cross veins are, however, primarily structural reinforcements, and in some cases, they are organized in a way to facilitate special bending paths. The veins are denser and thicker towards the LE and the wing base to support the structure against bending stresses and to reduce the inertial power requirement of the flapping by making the wing lighter towards the tip (Ennos 1989). Interestingly, this complex venation network of insect wings does not significantly affect average bending stiffness that is determined primarily by wing size, but venation affects the regional stiffness distributions to adjust the wing morphology (Combes 2010; Combes & Daniel 2003b).

A number of other specialized structural features of insect wings are also noteworthy. Flexion lines are essentially two-way joints or one-way hinges built in the wing structure facilitating chordwise bending and twisting (Wootton

1979, 1992). Some wings possess creases that do not normally bend during flight but that can prevent damage by crumpling reversibly during collisions, for example, near the tips of cranefly (*Tipulidae*) wings (Combes 2010). Fold lines, on the other hand, primarily function not during flight, but as a structural element making wing folding (e.g. beetles) possible when the wing is in rest (Brackenbury 1994; Combes 2010; Haas & Wootton 1996). The pterostigma of dragonflies, craneflies and bees is a pigmented spot with greater mass than the surrounding cuticle, which helps balance the chordwise distribution of mass and regulates wing pitch during flight (Åke Norberg 1972; Rajabi et al. 2016). Finally, wing corrugations strengthen the wing in the spanwise direction and can trap tiny vortices in their valleys that alter the effective wing cross section without additional mass (Combes 2010; Lian et al. 2014; Luo & Sun 2005; Shyy et al. 2013; Shyy et al. 2008). Finally, among insects we can also find a number of surface structures the functions of which are less understood (Combes 2010; Ghiradella 1998). Butterfly wings are covered with overlapping scales (Ghiradella 1994; Vukusic & Sambles 2001), true bugs have cone-shaped protrusions and small hairs cover the wings of flies (Ghiradella 1998), and small spines can be found on dragonfly wing veins (Combes 2010; Wootton 1992).

We first consider the chordwise wing flexibility and fluid-structure interaction in a 2D flow field, neglecting 3D flow effects. Three-dimensional effects, such as spanwise flow that seem to stabilize the LEVs (Birch & Dickinson 2001) or LEV-TiV interaction (Shyy et al. 2009), are noticeable in general. However, at smaller insect scales of $Re = O(10^2)$, the effects of spanwise flow on the overall aerodynamics are less important than at higher Reynolds numbers (Birch & Dickinson 2001; Shyy & Liu 2007). Also, the characteristics of the LEVs in two dimensions for plunging motions are representative of 3D flapping wings as long as the stroke-to-chord ratio is within the range of typical insects – that is around 4 to 5 (Alben & Shelley 2005; Rival et al. 2011; Wang 2005; Wang et al. 2004). Effects of 3D flow and spanwise wing flexibility are considered in Sections 2.5 and 2.6.

Consider an insect wing approximated with an elastic flat plate (Combes & Daniel 2003b) described by the plate equation (Eq. 1.17) or by a beam equation (Kang et al. 2011; Kang & Shyy 2013). The dimensional analysis of the plate equation leads to a set of non-dimensional parameters (e.g. the density ratio (Eq. 1.18), reduced frequency (Eq. 1.13 and Eq. 1.14) and Reynolds number (Eq. 1.9) as discussed in Section 1). Furthermore, we can define the thickness ratio $h_s^* = t_w/c$, frequency ratio $f/f_1 = 2\pi f/[k_1^2\{Eh_s^{*2}/(12\rho_f c^2)\}^{1/2}]$, for example, for a beam with k_1 being the first wave number of a cantilevered beam, and coefficient of lift $C_L = L/(0.5\rho_f U^2 c)$, where '$L$' is the lift force per unit length, and the first natural frequency, 'f_1', is measured in the chordwise direction between the LE and the TE.

Based on high-fidelity numerical computations of a wide range of wing motion, wing structure and geometry, and surrounding fluid (Kang et al. 2011; Kang & Shyy 2013) found that a relative shape deformation parameter γ plays a pivotal role in characterizing the force and propulsion characteristics of a flapping, flexible wing. Specifically, γ is defined as:

$$\gamma = \frac{2}{k} \frac{1 + \pi/(4\rho^* h_s^*)}{(f_1/f)^2 - 1} \tag{2.1},$$

which includes the effects of f/f_1, k, h_s^* and ρ^*, such that the individual effects of these materials and flapping parameters are integrated into a single parameter. When a plunge motion is imposed on the LE of the wing, the resulting wing camber deformations can be regarded as a passive pitch rotation $\alpha(t^*)$, the angle between the TE and LE. This pitch angle is purely passive due to the dynamic balance between the aerodynamic loading, elastic restoring force and inertia of the wing (Shelley & Zhang 2011). The parameter γ includes all three – that is inertia, stiffness and aerodynamic force terms. Kang and Shyy (2014) showed that the scaling relationship between the time-averaged force and γ is consistent with a scaling that was based on an experiment in air (Ramananarivo et al. 2011; Thiria & Godoy-Diana 2010). The scaling is also consistent with experiments performed in water (Dewey et al. 2013), where a frequency ratio was empirically determined by locating the maximum wingtip deflection.

In order to elicit the effects of the non-linear transient dynamics, we also consider an approximation of α with a first-order harmonic (FH) as $\alpha_{FH} = 90 - \alpha_a \cos(2\pi ft + \varphi)$, where the angular amplitude α_a and the phase lag φ between the plunge and angular motions are calculated by measuring α_m and α_e (Figure 2.1),

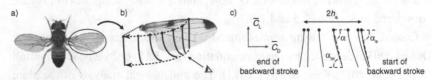

Figure 2.1 Schematic of the abstraction of the wing motion. (a) Top view of a *Drosophila melanogaster*. (b) Front view of a *Drosophila melanogaster* wing and schematics of two-dimensional cross sections of the wing, indicated by red solid lines. (c) Two-dimensional model of the flapping stroke shown in (b). Wing deformations lead to a passive pitch angle of α. Adapted from Kang and Shyy (2013).

the angles at the middle and end of the strokes, respectively. A scaling relationship exists (Kang et al. 2011; Kang & Shyy 2013) between the time-averaged lift coefficient and γ as:

$$\overline{C}_L^* = \frac{\overline{C}_L}{kh_s^*(f_1/f)^2\{1-(f/f_1)\}^{\sim}\gamma^a} \tag{2.2},$$

where a decreases from 2.0 for $\gamma < 1$ to 0.34 for $\gamma > 2$. This scaling is derived by considering the non-dimensional energy balance. When the wing deformations become more significant at $\gamma > 0$, the correlation worsens. This indicates that other factors may start to play a role as Eq. (2.2) only considers FH as insects usually flap with $f/f_1 < 1$ (Ramananarivo et al. 2011), and assumes aerodynamics forces from a high-frequency, small-deformation motion (Kang et al. 2011). When f/f_1 is small, the elastic, restoring effects dominate over the inertia effects, implying the work done on the wing is mostly in balance with the restoring elastic force. As f/f_1 increases, the influence from the wing inertia cannot be neglected anymore. Both the elastic-restoring and the inertial effects need to be considered.

The outcome is that the time-averaged lift coefficient versus γ covers a wide variety of wing flexibility and flapping characteristics relevant to insects as well as artificially devised flexible wings (Figure 2.2). Moreover, γ can be interpreted as a non-dimensional TE displacement relative to the LE. The chordwise wing

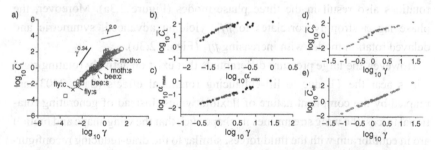

Figure 2.2 γ-scaling relationships. (a) Scaling given by Eq. (2.2) of the normalized lift $\overline{C}_L^*$ by γ for various data including insects Kang et al. (2011). (b,c) Correlations between $\overline{C}_L^*$, maximum angle α_{max}^* normalized in the same manner as for $\overline{C}_L^*$ and γ. (d,e) Normalized aerodynamic performance against γ from the current computations for (d) time-averaged power required $\overline{C}_P^*$ and (e) propulsive efficiency $\overline{C}_{eff}^*$. In (b,c), the colours indicate advanced ($\varphi > 95°$, blue), symmetric ($85 < \varphi < 95°$, red) and delayed ($\varphi < 85°$, black) rotation modes. Adapted from Kang and Shyy (2013).

deformations act as a passive pitch with an AoA α, which is a key measure of the aerodynamics and directly affects lift (Figure 2.2(b,c)). Furthermore, Figure 2.2 (d,e) shows that the normalized power required (Kang et al. 2011) and the propulsive efficiency defined as the ratio between the lift and the power required for the current cases scale with γ.

From Eq. (2.2), it follows that optimal lift is obtained when $f/f_1 = 0.5$ with other parameters fixed. For a self-propelled insect model (Ramananarivo et al. 2011), the best performance was observed at $f/f_1 = 0.7$, and the performance decreased by a factor of more than 4 at $f/f_1 = 1$ consistent with the current prediction. The current discussion also suggests that the maximum lift may be obtained at a frequency ratio well below the resonance condition because increasing the TE deformation relative to the LE results in delayed rotation mode, deteriorating the lift generation.

2.2 Optimal Passive Wing Rotation

When actively rotating, rigid wings exert force on the surrounding fluid, and the fluid adds to or subtracts from the lift due to translational mechanisms (Dickinson et al. 1999; Sane 2003). The benefits of the rotational mechanism depend on the phase difference between translation and rotation (Dickinson et al. 1999). Previous studies using rigid wings in hover have shown that the advanced rotation is optimal for low-AR wings (Dickinson et al. 1999; Trizila et al. 2011).

Depending on the wing structure and imposed motion, the passive wing rotations also result in the three phase modes (Figure 2.3a). Moreover, the phase lag φ strongly correlates to f/f_1, yielding advanced, symmetric and delayed rotation modes with increasing f/f_1 (Figure 2.3b).

However, the large pressure differentials that exist on actively rotating rigid wings near the TE due to lift-enhancing rotational effects (Sane 2003) are relaxed by the compliant nature of flexible wings. Instead of generating rotational forces, the wing streamlines its shape such that the wing shape and motion are in equilibrium with the fluid forces, similar to the drag-reducing reconfiguration of flexible bodies (Alben et al. 2002). Passive rotational angle α is purely due to deformation. The amplitude and the phase of the passive rotational angle scale with γ and f/f_1.

Moreover, α_m is measured at the instant of maximum translational velocity and is therefore indicative of the forces related to the translational mechanisms (Dickinson et al. 1999). For advanced rotations ($f/f_1 < 0.25$), α_m is greater than 70°. The wing orientation is almost vertical, producing little lift. On the other hand, deformations can become substantial for delayed rotation modes with $f/f_1 > 0.4$,

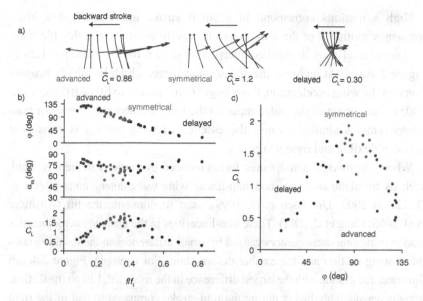

Figure 2.3 Aerodynamic forces and resulting passive wing motion generated by a flexible wing. (a) Schematics illustrating the three phase modes: advanced ($k = 0.95$; $\gamma = 0.58$), symmetric ($k = 0.6$; $\gamma = 2.59$) and delayed ($k = 2.0$; $\gamma = 24.9$). (b) Phase lag φ, mid stroke angle of attack α_m, and as a function of f/f_1. (c) As a function of φ. Adapted from Kang and Shyy (2013).

but since the translation is out of phase with the passive rotation, the resulting α_m and hence the lift remain smaller than for symmetric rotations. Figure 2.3c also shows that the resulting lift for the delayed rotations can be higher than for the advanced rotation modes, unlike the actively rotating rigid wings in hover (Dickinson et al. 1999). The translational forces peak when α_m is around 45 to 50° (Dickinson et al. 1999), which corresponds to $0.25 < f/f_1 < 0.4$. The resulting lift is of similar magnitude to that of rigid wings with active rotations (Dickinson et al. 1999). However, the lift is now optimal for symmetric rotation modes. This is also observed in the wing kinematics of FFs (Fry et al. 2005) and honeybees (Altshuler et al. 2005).

2.3 Lift Enhancement Due to Wing Streamlining

To further assess the effects of flexibility, Kang and Shyy (2013) compared the lift generation on a flexible wing to the lift on a rigid wing rotating with α_{FH}. As γ increased, wing camber deformations became larger, eventually causing the flexible wing performance to deviate from its rigid counterparts. For most of the considered motions, most flexible wings yield higher lift.

High k motions correspond to a small stroke amplitude and a high-frequency motion. For these motions, the resulting lift on the flexible wing is close to the lift on its rigid counterpart as shown for $k = 3.05$, $\gamma = 1.23$ in Figure 2.4(a,b). At higher k, the added mass force, which is linearly proportional to the wing acceleration, has a larger contribution to lift (Altshuler et al. 2005). The impact of the added mass on the total aerodynamic force for both wings remains similar because the essence of wing motion is consistent between flexible and rigid wings.

When k is lowered, aerodynamic forces induced by vorticity in the flow field, such as translational forces and non-linear wing-wake interactions (Birch & Dickinson 2003; Dickinson et al. 1999), start to dominate the lift (Altshuler et al. 2005; Kang et al. 2011). These non-linearities in the unsteady aerodynamics lead to intriguing consequences caused by small differences in the resulting non-linear wing motion and the camber deformation. For example, Figure 2.4(c,d) illustrates the motion with the largest difference in the mean lift. Lift on the flexible wing is considerably higher during the mid-stroke compared to that of the rigid wing, such that the lift generated by the flexible wing is superior: 1.2 versus 0.88.

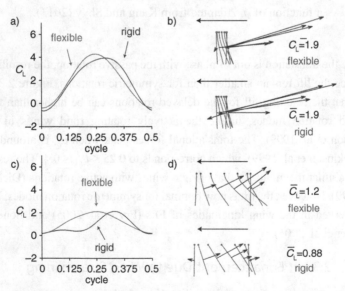

Figure 2.4 Difference in lift between flexible and rigid wings, rotating with α_{FH}. Lift time histories and wing shape schematics of the smallest and largest differences in lift are shown in (a,b) and (c,d), respectively. The integrated aerodynamic forces are indicated with the blue arrow. Adapted from Kang and Shyy (2013).

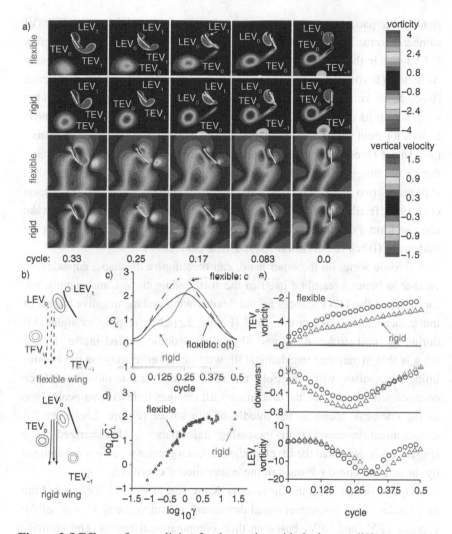

Figure 2.5 Effects of streamlining for the motion with the largest difference in lift (also shown in Figure 2.4(c,d)), compared to the flow around the rigid wing rotating with α. (a) Representative vorticity and vertical velocity fields based on computational fluid dynamics. (b) Schematic illustration of the wing-wake interaction. The subscript numbers indicate the stroke at which the vortices are shed: current (1), previous (0) and two strokes ago (−1). The arrows between the LEV_0 and TEV_0 illustrate the downwash. (c) Time history of lift normalized by c and $c(t^*)$. (d) as a function of γ. (e) Time histories of the vorticity of TEV_0, downwash and LEV_1. Adapted from Kang and Shyy (2013).

The wing-wake interaction provides an important mechanism responsible for lift enhancement over a flexible wing compared to its rigid counterpart, as highlighted in Figure 2.5. Here the unsteady flow field of a flexible wing and its rigid counterpart, rotating with α, are considered to focus solely on the effects of camber deformations (Kang & Shyy 2013). The LEV and the trailing-edge vortex (TEV) shed in the previous motion stroke, denoted as LEV_0 and TEV_0, respectively, form a vortex pair and induce a downward wake around the mid-stroke (Birch & Dickinson 2003; Trizila et al. 2011), which in turn interacts with the wing during its return stroke. In general, the outcome of the wing-wake interactions for rigid wings vary depending on the wing kinematics and wake structure. Under favourable conditions, added momentum causes the lift to increase during the first portion of the stroke (Dickinson et al. 1999). The wing then experiences two peaks in lift: the wake capture and the delayed stall (Dickinson et al. 1999; Trizila et al. 2011) peaks as shown in Figure 2.5e. However, the wake can also form a downward flow, causing the lift to drop significantly during the mid-stroke (Birch & Dickinson 2003; Trizila et al. 2011).

A flexible wing, on the other hand, can reconfigure its shape, adjusting its camber to better streamline itself to the surrounding flow. Consequently, the formation of the TEVs reduces for the flexible wing and the negative impact of the induced downwash is mitigated (Figure 2.5(a,b)), leading to higher lift during the mid-stroke. Kang and Shyy (2013) demonstrated that a flexible wing is able to generate translational lift with higher efficiency via the streamlining mechanism, while at stroke reversals, it lessens rotation-related force enhancement. As a result, the time history lift changes from the two-peak shape to the one-peak shape in the middle of the stroke (Figure 2.5c). The lift enhancement increases with increasing deformations, characterized by γ (Figure 2.5d). Note that the lift coefficient is even greater when it is normalized by the original chord c instead of the instantaneous chord $c(t)$.

To quantify this streamlining process, Kang and Shyy (2013) estimated the magnitudes of the non-dimensional downwash v_d and vorticity $\nabla^* \times \mathbf{u}^*$ of the vortices TEV_0 and LEV_1 based on their computational results. The vorticity magnitude of TEV_0 for the rigid wing is higher, which indeed correlates to a stronger downwash and a higher induced AoA. An induced AoA is the flow velocity direction in the perspective of the wing. On the other hand, the stronger downwash delays the increase of the LEV for the rigid wing.

2.4 Aeroelasticity in Three-Dimensional Flow Fields

Previous sections show that the lift was found to strongly depend on the instantaneous AoA, resulting from passive pitch associated with wing deformation. Other studies have also demonstrated the effect of flexibility on the improvement of the propulsive force and the propulsive efficiency at large AoA, normally greater than 60° (Heathcote & Gursul 2007; Kang et al. 2011; Kang & Shyy 2013, 2014; Percin et al. 2011; Shyy et al. 2010; Wu et al. 2011;

Zhao et al. 2010). One key mechanism is that the rear deformation results in an effective projected area for the propulsive forces to develop (Shyy et al. 2010).

These effects are, however, 2D in which the streamwise flow interacts with the chordwise flexibility. Further insight into the fluid-structure interactions in a 3D context is needed. In this regard, Fu et al. (2018) investigated the flexibility effects on the aerodynamic performance of isotropic wings rotating more than 120° at an intermediate AoA = 45° and Reynolds number $Re = 5.3 \times 10^3$, matching medium-sized insects and practical MAV designs. Although the AoA of a flapping wing often varies rapidly during stroke reversal, the AoA change is small during the majority of the flapping period of some flight missions or patterns (Shyy et al. 2010). Wings with a stiffened LE were studied over a vast stiffness range and for two *AR*s. A broad range of wing flexibilities – namely flexible (F1 and F2), very flexible (V1 and V2) and rigid wings (R1, R2 and R3) – were investigated and compared. Wing rotation was measured with the rotational angle φ.

The time course of the lift of certain flexible wings is in general similar to that of rigid wings (Figure 2.6). Flexible wings, especially F2, outperform the rigid counterparts in lift in two stages: the initial stage of the motion SL1 when there is large acceleration and in the second half of the period SL3. The impulse of the lift force in SL1 lasts longer for flexible wings than that for rigid cases, providing more lift. During the second half of the period, the wing is decelerating and being restored to its original shape. The restoration process may unleash the elastic energy stored in the wing by deformation and thus enhance lift. Also, during the deceleration phase, the deformation of flexible wings may help to keep the LEV stable while on rigid wings the outboard LEV breaks down during this stage. The lift enhancement in these two stages counterbalances the slight lift decrease in SL2, making flexible wings perform similarly to or better than rigid wings in lift.

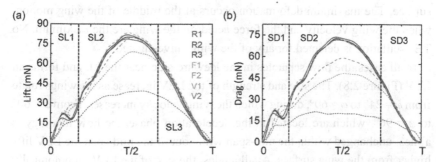

Figure 2.6 Time history of (a) lift and (b) drag forces in one period. Different wing models are in different colours, where R1, R2 and R3 denote rigid, F1 and F2 flexible, and V1 and V2 very flexible wings. Adapted from Fu et al. (2018).

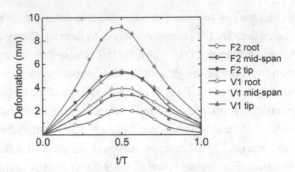

Figure 2.7 Wing deformation histories at the root, the mid-span and the tip for F2 (flexible) and V1 (very flexible) over the whole period of the motion. Adapted from Fu et al. (2018).

By contrast, no obvious drag increase is shown in the first or third stages, SD1 and SD3, during acceleration and deceleration, respectively. The pronounced decrease in drag in stage SD2 makes the total drag much smaller than in the rigid cases, leading to improved efficiency. The difference in the sensitivity of lift and drag forces to the change in the AoA due to wing deformation is associated with the reorientation of the total force (Shyy et al. 2010). For very flexible wings, both the lift and the drag decrease rapidly with increasing flexibility over the whole period. In addition, due to the wing deformation, the force peaks of the flexible and the very flexible wings shift slightly aft compared to the rigid cases.

Figure 2.7 shows the course of the deformation with the wing motion at the root, the mid-span and the tip for F2 and V1. The wings deform in the first half of the motion and are gradually restored to their original shape in the second half. The deformation history is in accordance with the force change on the wing surface. The maximum deformation occurs at the middle of the wing motion, when the wing velocity and the force acted on the wing are the maximum. No TE oscillation is detected for any of the wings investigated.

In all cases, the flow separates in the leeward side near the LE and forms an LEV (Figure 2.8). The size and strength of the LEV increase as the wings rotate from $\varphi = 24°$ to $\varphi = 60°$, during which the wing velocity increases. From $\varphi = 60°$ to $\varphi = 96°$, which are located in the deceleration phase, the flow topology is almost unchanged within the mid-span while out of the mid-span the LEV lifts higher from the wing surface. At all angles, the size of the LEV monotonically increases along the span up to near the wingtip. Near the wingtip, the LEV interacts with the TiV and the interaction shrinks its size (Zhou et al. 1999). The vortex structure around the flexible wing F2 is similar to that around the rigid wing R1 in the inner span and more stable than that around R1 in the outer span.

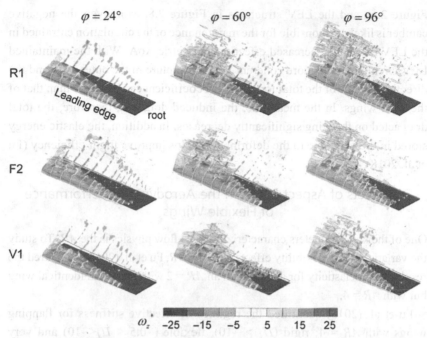

Figure 2.8 Dimensionless vorticity distribution ω_z^* at $\varphi = 24°$ (first half of the motion), $\varphi = 60°$ (middle of the motion) and $\varphi = 96°$ (second half of the motion) around wings of R1 (rigid), F2 (flexible) and V1 (very flexible) from particle image velocity measurements. Adapted from Fu et al. (2018).

However, the whole vortex structure around the very flexible wing V1 is evidently weaker. This was supported by the evolution of the circulation of the LEV around each wing.

The induced AoA also changes with varied flexibility. The mean-induced AoA at the mid-span during $\varphi = 12°$ and $\varphi = 108°$ was 22.8 °, 21.4°, 20.4 ° and 16.9° for R1, F1, F2 and V1, respectively, decreasing with increasing flexibility. This change in the induced AoA led to a drop in induced drag in accordance with the tilting in the total force acted on the wing.

With the deformation of the flexible wings, a small negative camber with a maximum camber deformation of no more than 1.5% of the chord forms passively. The effective geometric AoA, which is the AoA including the camber deformation, varies along the span, leading to twisting in the spanwise direction. Fu et al. (2018) show that the twisting helps maintain high AoA in the inboard region of the wingspan where the LEV is stable and decreases the AoA in the outboard region where the LEV tends to break down. The twisting also provides a stabilizing influence on the LEV in the outboard region during the deceleration phase and enhances lift (Nakata & Liu 2012), as illustrated in

Figure 2.6a and the LEV structures in Figure 2.8. Moreover, the negative camber is likely responsible for the maintenance of the circulation entrained in the LEV under the decreased effective geometric AoA. With the maintained LEV circulation, the more coherent vortex structure in the outer span and the direction change of the total force, the lift coefficient can be larger than that of the rigid wings. In the meantime, the induced drag and, therefore, the total drag acted on the wing significantly decreases. In addition, the elastic energy stored in the wing due to the deformation helps improve flight efficiency (Fu et al. 2018).

2.5 Effects of Aspect Ratio on the Aerodynamic Performance of Flexible Wings

One of the key parameters characterizing a 3D flow physics is the AR. To study the variation of the flexibility effects with the AR, Fu et al. (2018) compared the resulting aeroelasticity for a wing with an $AR = 2$ with that of an identical wing but with $AR = 4$.

Fu et al. (2018) quantified the regimes of effective stiffness for flapping wings with $AR = 4$: rigid ($\Pi_1 > {\sim}10$), flexible (${\sim}0.5 < \Pi_1 < {\sim}10$) and very flexible ($\Pi_1 < {\sim}0.5$) based on the aerodynamic performance. In the rigid regime, both the lift and drag coefficients show little change with varied effective stiffness Π_1 as shown in Figure 2.9. In the flexible regime, the wings experience a pronounced decrease in drag while maintaining a similar or even enhanced lift compared to their rigid counterparts. In the very flexible regime, both the lift and the drag coefficients decease severely with increasing compliance of the wing. Compliant wings, covering flexible wings and very

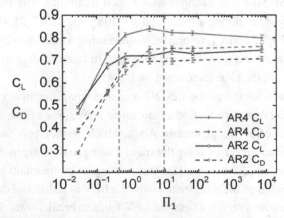

Figure 2.9 Lift and drag coefficients for wings with $AR = 2$ and $AR = 4$. Adapted from Fu et al. (2018).

flexible wings, improve the lift-to-drag ratio through predominant reduction in drag with little compromise in lift. Very flexible wings can also be adopted to further reduce the drag on the premise that the corresponding lift is sufficient to support the weight.

However, the three stiffness regimes located for $AR4$ wings cannot apply to $AR2$ cases. For the lower-AR wings, the rigid range extends to much smaller Π_1 with a value of around 0.6. With Π_1 further decreasing, both the lift and drag coefficients rapidly drop. No clear flexible regime was found for this low-AR case. While a flexible regime may also exist, the range of this regime for the low-AR case is much narrower. Therefore, wings with a high AR benefit from a wider flexible regime.

Incorporating the Reynolds number, the flexural stiffness and the AR, the effective stiffness can be rewritten as:

$$\Pi_1 = \frac{\rho_f E I_c}{\mu^2 c^2 Re^2 AR} \tag{2.3},$$

where EI_c is the chordwise flexural stiffness. For the same fluid type, the density and the dynamic viscosity can be considered as constant. In addition, Combes and Daniel (2003b) estimated that the chordwise flexural stiffness scales with the square of the chord length. With this estimation, one can infer that Π_1 is approximately proportional to the reciprocal of the product of Re^2 and AR, with Re characterizing the kinematics of the wing and AR indicating the wing morphology.

Although the flexibility effects on the aerodynamic performance are strongly correlated with the AR, the correlation has not received sufficient attention in previous studies. As shown in our current results, a compliant wing always exhibits lower lift than its rigid counterpart for $AR = 2$ while for $AR = 4$, a compliant wing can exhibit higher lift. The correlation between the flexibility and the AR may explain the controversy in literature. For instance, results by Zhao et al. (2010) on flapping wings with $AR = 3$ revealed no intermediate wing flexibility that could optimize aerodynamic performance, which is different from the results on wings with $AR = 4$, but similar to the results for $AR = 2$. The correlation between the flexibility effects and the AR are also likely to change with the wing kinematics. For instance, Ryu et al. (2016) studied the aerodynamic force on rigid and flexible wings with $AR = 3$ under a combined rotating and pitching motion. They found greater aerodynamic force on a flexible wing than a rigid wing. While the vorticity of the LEV and the corresponding lift decreased, Ryu et al. (2016) found that the greater aerodynamic force was caused by LEV behaviour shortly after the wing reversal.

The correlation between the flexibility effects and the AR is associated with the twisting along the span. For the high-AR case of $AR4$, the twisting helps stabilize the LEV structure in the outer span during the deceleration phase and enhances lift (Figures 2.6a and 2.8). By contrast, for the low-AR case of $AR2$, the LEV structure stays coherent in the outer span. The twisting plays no role in stabilizing the outer LEV but decreases the LEV structure. It is seen that compliant wings with a high AR have a flexible regime, unlike low-AR wings.

In the very flexible regime, the drag further decreases, but the wings also experience severe deterioration in lift. In this regime, the large deformation on the wing surface strongly reduces the effective geometric AoA. The AoA decreases up to 7° and the camber rate is up to 3.7%. Accordingly, the circulation entrained in the LEV decreases significantly. With the decrease in the LEV circulation, the lift cannot be maintained on very flexible wings. However, in all cases, the LEV core trajectories are found to be insensitive to flexibility. This implies that the LEV stays closer to the wing surface for compliant wings, having the potential to result in increased suction during flight (Percin et al. 2011).

2.6 Summary and Concluding Remarks

When actively rotating rigid wings exert force on the surrounding fluid, the benefits of the rotational mechanism depend on the phase difference between translation and rotation. It is observed that the advanced rotation is optimal for rigid low-AR wings in hover.

However, the large pressure differentials that exist on actively rotating rigid wings near the TE due to lift-enhancing rotational effects can be relaxed by the compliant nature of flexible wings. Instead of generating rotational forces, the wing streamlines its shape such that the wing shape and motion are in equilibrium with the fluid forces. As a consequence, the lift is now optimal for symmetric rotation modes. Overall, the wing-wake interaction provides an important mechanism responsible for lift enhancement over a flexible wing compared to its rigid counterpart.

For a flexible wing, twisting helps keep a high AoA in the inboard region of the wingspan where the LEV is stable and decreasing the AoA in the outboard region where the LEV tends to break down. The negative camber of the flexible wing maintains the circulation entrained in the LEV under the decreased geometric AoA. With the maintained LEV circulation and a more coherent vortex structure in the outer span, the lift coefficient can be larger than that of a rigid wing. In the meantime, the induced drag and, therefore, the total drag acting on the wing significantly decreases. In addition, the elastic energy stored in the wing due to the deformation helps improve aerodynamic efficiency.

Furthermore, a flexible wing is more resistant to impacts of any kind, which greatly helps the survivability of a living creature or an aircraft. Certainly, there are areas where flexible wing aeroelasticity needs further research. Insect wings are inherently anisotropic structures with venations and corrugations that result in a complex distribution of flexibility variations. Many insects generate more lift in the downstroke than in the upstroke and the wings are more rigid in the downstroke (Hedrick et al. 2015). This implies that insect wings may operate in the flexible regime in the downstroke and in the very flexible regime in the upstroke. The effective stiffness and the flexibility effects of insect wings are also likely to change with flight patterns. The characterization of such complex natural structures is still in its infancy but expected to show progress in the near future. As a first step, recent works address the fluid dynamic and propulsive characteristics of flapping foils that show a gradual change in their flexibility (Lucas et al. 2015; Riggs et al. 2010; Salumäe & Kruusmaa 2011).

3 Stability and Dynamics of Flexible Flapping Wings

Many flying insects can hover, which is an energetically demanding flight mode. In the low-Reynolds-number flow regime, insects use the well-known unsteady lift enhancement mechanisms of delayed stall, rotational lift, added mass forces and wake capture to produce the required lift to stay aloft (Shyy et al. 2013).

Despite the advancements in our understanding of insect aerodynamics, the development of bio-inspired small robotic flyers has faced many challenges. One such challenge is understanding the flight dynamics of insects, particularly in hover. Insects not only generate sufficient aerodynamic forces to sustain their weight and overcome drag in hover, they also must hold their position in equilibrium. In general, the stability of insects is described by presenting the eigenvalues of the linearized system matrix (Cheng & Deng 2011; Sun 2014; Taha et al. 2012). The real part of an eigenvalue corresponds to the rate of growth (Bluman & Kang 2017b) and the imaginary part corresponds to the natural frequency of any oscillatory motion. If any eigenvalue has a positive rate of growth, then the system is unstable. The literature reports on hover eigenvalues for a number of insect families, including drone flies (DFs), fruit flies, bumblebees (BBs), hawkmoths and hover flies. These studies include several different aerodynamic modelling techniques of varying fidelity (Bluman & Kang 2017b; Cheng & Deng 2011; Faruque & Humbert 2010; Liang & Sun 2013; Orlowski & Girard 2012b; Taha et al. 2015). Yet they overwhelmingly share the same natural modes of motion: the hovering equilibria are reported to be unstable for flapping wing flyers (Orlowski & Girard 2012a; Sun 2014).

However, none of these studies have considered flexible wings (Shyy et al. 2016; Sun 2014; Taha et al. 2012) in spite of the fact that insects themselves feature flexible wings (Altshuler et al. 2005; Ennos 1988; Fry et al. 2005; Shyy et al. 2013; Sunada et al. 1998; Young et al. 2009), and most flapping wing MAVs produced to date (Coleman & Benedict 2015; Desbiens et al. 2013; Keennon et al. 2012; Ma et al. 2013; Shang et al. 2009; Shyy et al. 2005; Tay et al. 2015; Tu et al. 2020b) also have flexible wings. Flexible wings produce lift differently from their rigid counterparts (Kang et al. 2011; Kang & Shyy 2013, 2014) by affecting the surrounding unsteady flow (Kang & Shyy 2013; Mountcastle & Combes 2013) and the efficiency and the timing of passive wing kinematics (Eldredge et al. 2010; Sridhar & Kang 2015). Flexible wings tend to require less power to flap because the passive deflection produces less drag and torque penalties associated with rotating the wings (Eldredge et al. 2010).

In this section, we review the recent advancements in the area of stability and dynamics of flexible flapping wings in hover. We first introduce the dynamic equations of motion for a flapping wing flyer and stability derivatives. Then we summarize the reports in the literature that the hover equilibrium is unstable under the assumption of rigid wings. However, a flexible wing passively deforms its wing shape in the presence of perturbations, which leads to qualitatively different hover stability characteristics. In particular, we discuss the recent study by Bluman et al. (2018) that states the hover equilibrium of flexible flapping wings is linearly stable. These findings suggest that the wing flexibility affects both the unsteady aerodynamics and the dynamic characteristics of flapping wings. With flexible wings, insects may passively stabilize their hover flight, which can benefit the designs of synthetic flapping wing robots.

3.1 Flight Dynamics of Flapping Flyer with Flexible Wings

The flight dynamics of a flapping flyer with flexible wings are governed by the complex three-way coupling of aerodynamics, structural dynamics and flight dynamics (Bluman et al. 2018; Bluman & Kang 2017a; Shyy et al. 2016; Taha et al. 2012). The aerodynamic force generation, governed by the non-linear unsteady viscous Navier-Stokes equations, depends on the wing shape deformation, which at the same time depends on the aerodynamic loads. Additionally, in free flight, body motion affects the fluid-structure interaction of flexible flapping wings, which in turn accelerates the body. Whereas hover is a given condition in most studies on flapping wing aerodynamics without the need for body motion, appropriate initial conditions and control inputs must be determined so that the flapping flyer achieves hover equilibrium prior to assessing its stability characteristics.

Most studies impose a sinusoidal function as the bio-inspired flapping motion ζ (Berman & Wang 2007) on the LE of a wing with a flapping frequency f and an amplitude Z (Figure 3.1). For hover, the reference velocity is a wing velocity at a representative point on the wing.

The key feature in the dynamics of flexible flapping wings is that the wing pitch angle is not prescribed as opposed to rigid wings, which require an active pitch rotation to efficiently produce lift and thrust as discussed in the literature (Shyy et al. 2013). Rather, the instantaneous wing pitch α_{flex} results from the dynamic balance between the wing's inertia, elastic restoring force and aerodynamic forces. Furthermore, body motion is superimposed on the flapping motion, both of which can affect the resulting wing motion. Unlike most rigid or flexible flapping aerodynamics studies (Shyy et al. 2010; Shyy et al. 2013), where the insect is fixed and only resulting forces are analysed, the body responds to the instantaneous aerodynamic forces and moments. As a result, the LE of the wing also moves through multiple degrees of freedom, which in turn affects the resulting aerodynamics and fluid-structure interaction. Mathematically, the dynamical system can be written as:

$$\frac{d\mathbf{x}}{dt} = \mathbf{f}(\mathbf{x}, t), \tag{3.1}$$

where $\mathbf{x}$ is the state vector representing the position and attitude of the body centre of gravity (CG) and their rates and t is the time. The equations of motion can be derived using different methods, and we refer to various studies for details (Bluman & Kang 2017a, 2017b; Orlowski & Girard 2012a; Sun 2014; Taha et al. 2012). Although the lateral dynamics and coupling between the lateral and longitudinal dynamics are important, we focus on the longitudinal dynamics and stability of insect models in relation to hover equilibrium. The longitudinal stability of hovering insects contains the largest source of instability

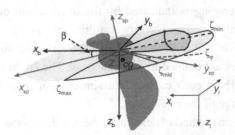

Figure 3.1 Fruit fly model with pertinent reference frames. Subscripts I, b and sp refer to the inertial, body and stroke plane reference frames, respectively. Adapted from Bluman et al. (2018).

(Sun 2014). Then the state vector reduces to $\mathbf{x} = [u, w, q, {}_I x_{cg}, {}_I z_{cg}, \theta]^T$, where u and w are the body velocities at the body CG along the x_b and z_b axes, respectively, and q is the pitch velocity about the y_b axis. The body CG positions ${}_I x_{cg}$ and ${}_I z_{cg}$ and the pitch attitude θ are in the inertial frame of reference (Figure 3.1).

To determine the stability derivatives, Eq. (3.1), which represents a set of non-linear, time-varying equations of motion, can be linearized as:

$$\frac{d\mathbf{x}}{dt} = A\mathbf{x} + B\mathbf{u}, \tag{3.2}$$

where A is the system matrix, B is the control input matrix and $\mathbf{u}$ is a control vector. As the aerodynamic forces and moments on the flapping wing flyer are independent of the body CG position, A is typically expressed as a 4×4 matrix (Sun 2014) for the open-loop dynamics of a flapping wing flyer as:

$$A = \begin{bmatrix} \frac{1}{m}X_u & \frac{1}{m}X_w & \frac{1}{m}X_q & g \\ \frac{1}{m}Z_u & \frac{1}{m}Z_w & \frac{1}{m}Z_q & 0 \\ \frac{1}{I_{yy}}M_u & \frac{1}{I_{yy}}M_w & \frac{1}{I_{yy}}M_q & 0 \\ 0 & 0 & 1 & 0 \end{bmatrix} \tag{3.3}$$

where the stability derivatives X_u/m, X_w/m, X_q/m, Z_u/m, Z_w/m, Z_q/m, M_u/I_{yy}, M_w/I_{yy} and M_q/I_{yy} describe the cycle-average response of a flyer in equilibrium to a perturbation in a particular degree of freedom. X, Z and M are the aerodynamic forces acting on the body CG in the x_b and z_b directions and pitch moment about the body CG, respectively. m is the total mass and I_{yy} is the body inertia about the y_b axis. The horizontal velocity damping X_u/m is defined as:

$$\frac{X_u}{m} = \frac{\Delta \dot{u}}{\delta u} \equiv \frac{\bar{\dot{u}}(u_0 + \delta u, w_0, q_0, \theta_0) - \bar{\dot{u}}(u_0, w_0, q_0, \theta_0)}{\delta u}, \tag{3.4}$$

where cycle-averaging is indicated by the overbar and δu is a horizontal velocity perturbation. The hover equilibrium is a solution where $\bar{\dot{u}} = \bar{\dot{w}} = \bar{\dot{q}} = \bar{u} = \bar{w} = \bar{q} = 0$ and is indicated with the subscript 0. Other stability derivatives are defined in a similar way. The control matrix B can be obtained by perturbing each control in a similar fashion and determining its effect on the average of the rate vector.

The most common method of describing the stability of insects is by presenting the eigenvalues λ_i of the linearized system matrix A (Cheng & Deng 2011; Sun 2014; Taha et al. 2012). An insect is said to be stable about its hover equilibrium if all eigenvalues, also known as open-loop poles, possess negative real parts. If at least one eigenvalue has a positive real part, the hover equilibrium becomes unstable.

The averaging approach in Eq. (3.4) assumes that the flapping timescale is much shorter than the body timescale (Taylor & Thomas 2002). This is the case for small insects such as flies and bees with a flapping frequency higher than 100 Hz and smooth flight trajectories (Taha et al. 2012). For larger insects and birds, the flapping frequency is much lower and the coupling between the flapping wing aerodynamics and body dynamics becomes more pronounced (Kang et al. 2018). In this case, the averaging approach may not be valid and more advanced approaches, such as the Floquet theory (Su & Cesnik 2011; Taha et al. 2012; Wu & Sun 2012) or non-linear approaches (Liang & Sun 2013) need to be used.

3.2 Stability of Hover for Rigid Wings

Figure 3.2 depicts the open-loop poles of the longitudinal dynamics of fruit flies (Bluman & Kang 2017b; Cheng & Deng 2011; Faruque & Humbert 2010), BBs (Sun & Xiong 2005) and DFs (Sun et al. 2007; Wu et al. 2009), reported in the literature. The wing kinematics are slightly different for each insect model, but rigid wings are assumed. Despite the variations in the wing kinematics and insect species, the qualitative response of all of these simulations is the same. Each insect has an unstable oscillatory mode, a fast stable subsidence mode and a slow stable subsidence mode. Because of the unstable mode, the hover equilibrium is unstable.

The primary reason for the instability of the rigid wing models is the large magnitude of the stability derivative M_u relative to the others (Cheng et al. 2011; Sun & Xiong 2005; Taha et al. 2014). M_u is the rate at which an insect tends to pitch up in the presence of a horizontal velocity perturbation. Although various sources of damping exist, this large pitch rate causes the insect to diverge away from hover. The rate depends on the insect configuration: it is about 29 times slower than the flapping frequency for hawkmoths and 114 times slower for hover flies (Taha et al. 2015).

Bluman and Kang (2017b) argue that the physical sources of the large stability derivative experienced by a rigid wing stem directly from several of the unique unsteady low-Reynolds-number force-production mechanisms: translational lift from delayed stall, rotational lift and wing-wake interaction. Each of these mechanisms destabilizes the hover equilibrium of an insect with rigid wings. The net rearward translational drag force, acting above the body CG, induces nose-up pitch (Sun & Xiong 2005). Also, the stronger rotational lift during stroke reversal from the advancing to retreating strokes causes a larger nose-up pitch momentum than in the opposite half stroke (Bluman & Kang

2017b; Cheng & Deng 2011). Finally, the wing-wake interaction, where the wing gains momentum from the near field vortex structures immediately following stroke reversal, creates a fore-to-aft lift imbalance and induces a nose-up tendency under a gust (Bluman & Kang 2017b).

Both the quasi-steady aerodynamics model and the Navier-Stokes equation model are used to calculate the eigenvalues shown in Figure 3.2. The magnitudes of the poles are larger when the Navier-Stokes equations are considered to model the unsteady viscous flow around the flapping wings. Bluman and Kang (2017b) demonstrated that the wing-wake interaction, which cannot be captured by a quasi-steady model, contributes to the instability. The wing-wake interaction increases the size of M_u, ultimately increasing the natural frequency of the response of the flapping wing flyer and decreasing its doubling time.

3.3 Passively Pitching Flexible Flapping Wings in Hover

To investigate the hover stability of flapping wings with flexible wings, Bluman et al. (2018) used an insect flight dynamics simulator that is fully coupled to the unsteady aerodynamics and flexible wing dynamics (Bluman & Kang 2017a) to calculate the stability derivatives. The detailed discussion of the governing equations, computational methodology, computational set-up and validation studies is reported in (Bluman & Kang 2017a). They varied the flapping wing amplitude Z, stroke plane angle β and stroke mean angle ζ_φ (Figure 3.1), so that the fruit fly model with a mass of about 1 mg is in a hover equilibrium (Bluman & Kang 2017a). In hover, the time-averaged lift offsets the weight and the time-averaged drag is zero.

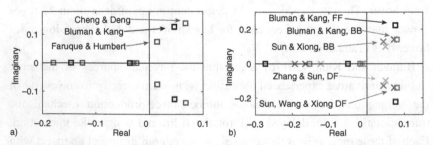

Figure 3.2 Open-loop poles for the longitudinal dynamics of a fruit fly using the (a) quasi-steady and (b) Navier-Stokes aerodynamic models, reported in the literature (Bluman & Kang 2017b; Cheng & Deng 2011; Faruque & Humbert 2010; Sun et al. 2007; Sun & Xiong 2005; Zhang & Sun 2010). FF: fruit fly; BB: bumblebee; DF: drone fly. Adapted from Bluman and Kang (2017b).

The flap amplitude that yields hover depends on the frequency ratio. Specifically, the flap amplitude is $Z = 80.5°$ for $f/f_1 = 0.330$ and decreases to $69.8°$ for the most flexible case ($f/f_1 = 0.461$). This range of flap amplitude agrees well with the amplitudes used by live fruit flies reported in the literature: $60° < Z < 76°$ (Fry et al. 2005; Hedrick et al. 2009; Lehmann & Dickinson 1997). The calculated stroke plane angle for hover was $\beta = -11°$ to $-13.4°$, which is close to the observed value of $12°$ (Fry et al. 2005) for fruit flies.

Thanks to the fluid-structure interaction implemented in the numerical model, the passive pitch angle α_{flex} variations resulting in hover equilibrium could also be determined. Figure 3.3b depicts the time history of pitch angle, experimentally observed for hovering fruit flies by Fry et al. (2005) in comparison with the values derived numerically. Although there are some differences due to modelling and experimental uncertainties, the resulting passive pitch motion from the numerical simulations resembles that of fruit flies in the numerical model of Bluman et al. (2018) and Bluman and Kang (2017a).

This passive pitch angle α_{flex} is in contrast to active wing rotations, but is required for rigid wings to produce sufficient lift to hover. Depending on the combination of the structural parameters and wing motion, passive pitch angles can be sufficient to generate enough lift to remain aloft (Kang & Shyy 2013; Sridhar & Kang 2015). The extent to which insects actively rotate their wings is not fully known. Although they possess the musculature to achieve pitch rotation (Bergou et al. 2007), experiments have also observed tip-to-root torsion waves and other evidence that insects typically rely on passive deformation of their wings to fly (Bergou et al. 2007; Ennos 1988).

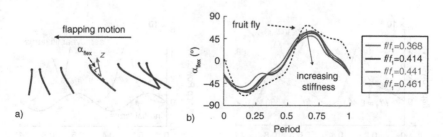

Figure 3.3 Passive pitch motion α_{flex} due to wing flexibility and body dynamics. (a) Schematic of the wing deformation for $f/f_1 = 0.41$. (b) Time histories of passive pitch angles resulting from changing the wing flexibility alongside observed kinematics from a fruit fly (Fry et al. 2005). Adapted from Bluman et al. (2018).

3.4 Power Required to Hover with Flexible Wings

A flapping flyer must expend energy to generate the power to hover, to move and to manoeuvre. For a wing with passive wing deformation, only flapping power is required. On the other hand, for rigid wing models, a pitch actuation mechanism is required to actively rotate the wing, which adds to the power required. Power required is an outcome of combined forces and moments acting on the wings and the body of the flyer as well as the velocity and acceleration of the motion. The interplay between the forces and moments and the non-linear response of the body in hover equilibrium is discussed in detail by Bluman et al. (2018) and Bluman and Kang (2017b). Here, we focus on the resulting power required and the resulting stability implications (Section 3.5).

The time histories of power are depicted in Figure 3.4 for various frequency ratios. The aerodynamic contribution is significantly larger than the inertial, since aerodynamic power scales as Z^3 and inertial power scales with Z^2, where Z is the flapping amplitude (Lehmann & Dickinson 1997). The time histories of power also show qualitative agreement with the trends for live fruit flies reported in Fry et al. (2005). Additionally, the aerodynamic power in Fry et al. (2005) was largest at the middle of each half-stroke, whereas their inertial power had larger peaks in the same locations, where flapping acceleration is highest.

To illustrate the influence of wing flexibility on power, Bluman and Kang (2017b) plotted the specific power to achieve hover for different values of frequency ratio (Figure 3.5a). The power required compares favourably to the ranges of power required for a fruit fly reported in the literature. For the flexible wing, the required power monotonically decreases as the wing becomes more flexible. The stiffest wing requires 38.5 W/kg of power while the most compliant wing requires 23.1 W/kg, a reduction of 66%.

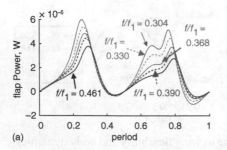
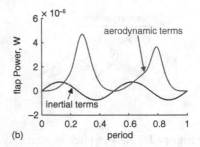

Figure 3.4 Time history of power required. (a) Time histories of flapping power for selected frequency ratios. (b) The time histories of the inertial (black) and aerodynamic (brown) contributions to flapping power for $f/f_1 = 0.390$. Adapted from Bluman and Kang (2017b).

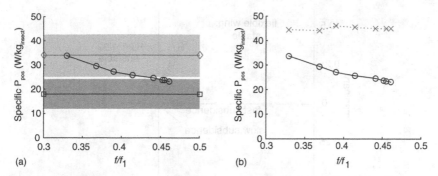

Figure 3.5 Specific power required. (a) The power required to hover for the flexible wing versus various frequency ratios (o) by Bluman and Kang (2017b). The required power of fruit flies from Lehmann and Dickinson (1997) (□) and Fry et al. (2005) (◊) are also plotted with the average values (solid) and the experimental range (shaded). (b) The power required for the flexible wing (o) compared to its rigid wing counterpart (×). Adapted from Bluman and Kang (2017b).

The power required to reach hover equilibrium for the flexible wing is much lower than the power required for a rigid wing using the pitch motion (Figure 3.5b). The rigid wing consistently requires approximately 45 W/kg of power to hover across a range of pitch amplitudes. The main reason for this difference is that the rigid wings shed more vorticity into the flow, which reduces lift in the wake valleys. The rigid wing needs a larger flapping amplitude to achieve hover as a result: the flapping amplitude for $f/f_1 = 0.414$ is $Z = 71.8°$; for the same pitch schedule, a rigid wing requires $Z = 84.6°$, which is an 18% increase. Since profile power increases with Z^3, the profile power is 60% larger for the rigid wing. Additionally, the inertial power due to flapping increases by 38%. Another reason is that active pitching power for rigid wings requires additional power. Both reasons demonstrate the benefit of utilizing flexible wings.

3.5 Flexible Wings Stabilize the Unstable Mode

Figure 3.6 shows that the real parts of all system eigenvalues associated with flexible wings are negative, suggesting that the hover equilibrium of flexible flapping wings is linearly stable. A flexible wing adjusts its structural dynamic response under a perturbation, a feature not available to a rigid wing. The enhanced damping associated with wing flexibility yields a stable hover equilibrium across a range of different wing stiffness values.

The hover equilibrium is unstable for a wide range of insect models with rigid wings. Bluman et al. (2018) demonstrated that, when wing flexibility is

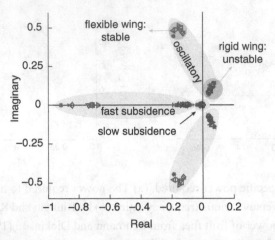

Figure 3.6 Open-loop poles for multiple insect species from studies that assume rigid wings (×: hover fly (Sun et al. 2007); ○: drone fly (Sun et al. 2007; Zhang & Sun 2010); +: bumblebee (Sun & Xiong 2005)) and flexible wings (green with different wing stiffness values) (Bluman & Kang 2017a). Adapted from Bluman et al. (2018).

considered, all system eigenvalues associated with flexible wings are negative (Figure 3.6), suggesting that the hover equilibrium of flexible flapping wings is linearly stable. The unstable oscillatory mode has become stable. The least stable mode is now the slow subsidence mode associated with the vertical motion of the insect, whereas the fast subsidence mode has become much more stable than for the rigid wings.

Bluman et al. (2018) further argue that only the stability derivatives Xu, Mu and Mq (Section 3.1) have a noticeable impact on the system poles. Similar to the rigid wing dynamics (Taha et al. 2014), increasing the magnitudes of Xu and Mq has a stabilizing influence on the oscillatory mode, which was the unstable mode for the rigid wings, and Mu is destabilizing.

Compared to the stability derivatives reported in the literature for rigid wings, the insect models with flexible wings are more sensitive to perturbations. Bluman et al. (2018) suggest that the main feature in the response of a flexible wing is that both the structural response and the surrounding unsteady flow respond to the imposed perturbation. The wing shape and motion remain the same for the rigid wing, and only the surrounding flow changes under a perturbation. This increased sensitivity does not necessarily mean, however, that the insect is less stable. On the contrary, the increased sensitivity significantly enhances the horizontal velocity damping $-Xu$ and the pitch rate damping $-Mq$, both of which stabilize the insect (Bluman et al. 2018).

According to Taha et al. (2014) and Bluman et al. (2018), the hover equilibrium can be stabilized if the horizontal velocity damping Xu and the pitch rate damping Mq are sufficiently large compared with Mu. For most rigid wing systems, Xu and Mq are too small compared to Mu to stabilize the hover equilibrium (Taha et al. 2014). For flexible wing systems, the ability for the wing to deform under perturbation provides the means to sufficiently increase the damping terms Xu and Mq.

Horizontal velocity damping Xu exists for both the rigid and flexible wings. Bluman et al. (2018) compared the interplay between the resulting wing motion and aerodynamics for hover and under perturbation for both the rigid and flexible wings. When the horizontal perturbation opposes the wing motion and the relative velocity of the wing is higher than at hover, horizontal rate damping results. One reason that Xu is larger for the flexible wing is that the flexible wing shape deformation responds to the horizontal perturbation, increasing its AoA. This increase in the AoA leads to a larger drag and damping. Furthermore, the ability for the flexible wing to deform in a qualitatively different way under a perturbation, the rate of wing shape deformation change, also results in a larger drag.

Bluman et al. (2018) found similar observations for the pitch rate damping Mq. The ability for the flexible wing to adjust its wing shape under the perturbation yields less lift in front of the body CG and larger drag.

3.6 Evolution of Flight from Hover

We have seen that open-loop poles of the flexible flapping wing flight dynamics system all have negative real components, suggesting that the hover equilibrium is stable. However, the stability derivatives were calculated after a linearization about the hover equilibrium of the averaged dynamics (Sun 2014; Taha et al. 2012). The real system is non-linear and time-varying. The lift production can be affected by the time history effects of the vortex dynamics and by the non-linear interaction with the structural response of the flexible wing. For example, a persistent downward jet sustained through a non-linear wing-wake interaction mechanism can decrease lift production (Trizila et al. 2011).

To test the hypothesis that wing flexibility passively stabilizes hover, Bluman et al. (2018) compared the longitudinal response of the flexible wing against two different rigid wing simulations: one with abstracted kinematics and the other with a pitch schedule that matches the flexible wing's passive pitch as shown in Figure 3.7.

Initially both the flexible and rigid wings are in hover. The periodic forcing of both wings causes cyclic variations in all three degrees of freedom. Eventually, both rigid wing systems diverge much more rapidly from the

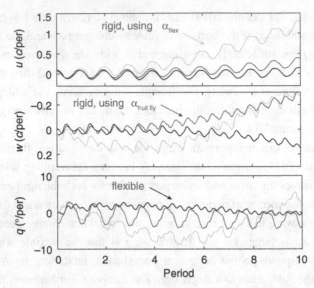

Figure 3.7 The longitudinal evolution of flight when released from hover for flexible and rigid wings, demonstrated by Bluman et al. (2018). The rigid wing results consist of a pitch motion from observations of hovering fruit flies (Fry et al. 2005) (red) and a motion with the pitch schedule based on the flexible wing response (blue). The flexible wing dynamics correspond to $f/f_1 = 0.41$ (green). Adapted from Bluman et al. (2018).

equilibrium than the flexible wing. The more rapid divergence of the rigid wings further demonstrates that the wing motion itself is not enough to stabilize a flapping wing flyer. Bluman et al. (2018) use these results to explain why the experimentally observed wing pitch kinematics of hovering fruit flies (Fry et al. 2005) do not confer any stability when simulated with rigid wings. When the pitching motion observed from an insect, which itself results from wing flexibility, is reproduced using a rigid wing simulation, the stability benefits are not seen. The true source of these benefits – that is the ability to adjust the wing shape and its dynamics under some perturbation – is not simulated.

3.7 Summary and Concluding Remarks

The stability in hovering as well as during transition from hovering to forward flight is particularly important for feeding (Liu et al. 2016), hunting (Combes et al. 2010; Windsor et al. 2014) and in evasive manoeuvres (Alexander 2002; Hedrick 2011) in nature, as well as practically relevant for MAVs having a wide range of mission roles (Shyy et al. 2013; Shyy et al. 2016). It was shown that rigid wings provide little or no stability. Flexible wings, however, react to perturbations with advantageous passive shape adaptations, making the hover equilibrium linearly stable. A flexible wing produces smooth passive pitch rotations that not only stabilize flight, but require less energy for wing actuation. On the other hand, many insects are physiologically capable of actively pitching their wings with the help of their wing root musculature (Bäumler et al. 2018; Büsse et al. 2015; Deora et al. 2017; Dickinson et al. 1993; Neville 1960), however, whether flying insects rely on active pitch control and to what extent is debatable (Bergou et al. 2007; Ennos 1988). Considering MAV flight control, the extent of active pitch necessary for stable flight with the least energy cost is yet to be clarified.

4 Aerodynamic Interactions of Tandem Winged Systems Based on Dragonflies

Dragonflies are among the most agile flying insects (Bode-Oke et al. 2018; Combes et al. 2012); they move each of their wings individually and change the interactions between forewings and hindwings instantaneously and continuously(Alexander 1984). Observations on free-flying dragonflies reveal that in-phase flapping generates high aerodynamic forces for quick, demanding manoeuvres or take off, while out of phase flapping is favoured for steady flight and hovering (Alexander 1984; Azuma & Watanabe 1988; Berman & Wang 2007; Norberg 1975; Rüppel 1989; Wakeling & Ellington 1997; Wang et al. 2003). Figure 4.1 shows the wings of a tethered dragonfly that executes flapping kinematics, typical for slow forward flight. In Figure 4.1, the course of the flapping positional angles of the wings and the geometric AoA at the mid-span are also given. The wing sections of 25%, 50% and 75% span are shown by painted white, yellow and white lines, respectively.

Previous investigations revealed essentially two types of interaction mechanisms between the forewing and the hindwing of dragonflies (Lian et al. 2014; Salami et al. 2019). When the forewing alters the incoming flow that is faced by the hindwing, it is defined as a downwash effect. On the other hand, the forewing also sheds complete vortical elements in its wake that result in inter-wing vortex interactions. A dragonfly's relatively long wing AR also results in highly complicated aerodynamic features along the spanwise direction.

Numerical (Lan & Sun 2001; Sun & Lan 2004; Wang & Russell 2007) and experimental (Hu & Deng 2014; Maybury & Lehmann 2004; Usherwood & Lehmann 2008) studies of hovering flight using either horizontal or inclined strokes have shown the downwash effect to be detrimental to vertical force generation. Downwash attenuates hindwing LEV circulation (Maybury & Lehmann 2004) in the case of forewing-led phasing. In the case of hindwing-

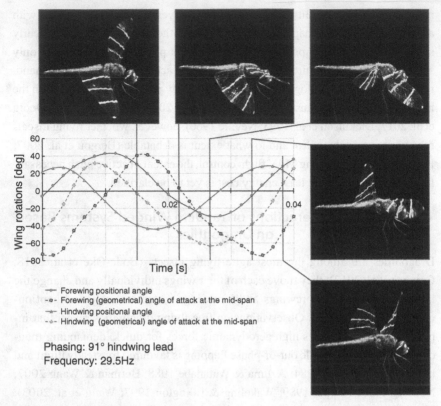

Figure 4.1 Typical wing kinematics of a dragonfly in slow forward flight mode (via tethered measurement). (Unpublished by authors).

led phasing, however, hovering with all four wings consumes 22% less power than hovering with just two wings (Usherwood & Lehmann 2008).

One of the key attributes of a flapping wing in generating lift is the delayed stall by the LEV, which can result in a substantial aerodynamic force for a flapping wing system (Dickinson et al. 1999; Dickinson & Gotz 1993; Ellington et al. 1996). The LEV generates a low-pressure area, which results in an increased suction force on the wing surface (Birch et al. 2004; Sane 2003; Shyy & Liu 2007). Direct vortex interactions could advance or delay LEV formation and shedding and, at the same time, strengthen or attenuate LEV circulation.

Maybury and Lehmann (2004) and Rival et al. (2011) experimentally revealed that a TEV shed by the forewing promotes the formation of the hindwing LEV in the case of 90° phasing of the wings. Hsieh et al. (2010) performed 2D numerical simulation and explained that the increase in hindwing thrust at 90° phasing is due to the fusion of the forewing downstroke TEV with the LEV of the upstroking hindwing. In a counter-stroking set-up, however, a similar vortex fusion amplifies the lift generated by the downstroking hind-wing (Hsieh et al. 2010). Furthermore, experimental (Hu & Deng 2014) and numerical (Xie & Huang 2015) studies showed that the forewing TEV and the hindwing LEV interact synergistically if they are formed in close proximity.

Additionally, Broering et al. (2012) concluded from their 2D numerical simulation of forward flight that at a system level, the tandem configuration produces a large thrust with a high propulsive efficiency but at the cost of lift efficiency when the wings are flapping in phase. Differently, a 90° or 180° phase flapping cycle greatly reduces the power consumption at the expense of thrust production (Broering et al. 2012). Zheng et al. (2015, 2016a, 2016b) studied the underlying interaction features of pitching and plunging wings that are either flexible or rigid while hovering (Zheng et al. 2016a), and both while hovering and in forward flight (Zheng et al. 2015, 2016b) at critical time instances when large force modulation occurred as a result of aerodynamic interactions. They also found synergy between the forewing TEV and the hindwing LEV when the two wings were flapping in phase. When the wings were flapping out of phase, the TEV shed by the forewing could either enhance the hindwing LEV (when their spins were opposite) or attenuate it (when their spins were the same) upon reaching the hindwing. They also showed that in some cases, the flow induced by the forewing LEV could increase the hindwing's effective AoA, resulting in an earlier LEV formation.

These previous investigations could uncover particular interaction scenarios that depend on the applied kinematics and geometrical set-up of the wings; however, less attention was paid to the spanwise variance of interaction features of the high-AR wings of dragonflies. Furthermore, previous parametric studies generally applied rigid wing modelling or simplified wing kinematics and the aerodynamic effects of changing the studied parameters were investigated in isolation. Dragonfly wings are highly complex, flexible structures (Combes 2010; Combes & Daniel 2003b; Sunada et al. 1998). Furthermore, what we see as the flapping of the wings is a sophisticated motion composed of three rotational motions (flapping, pitching and deviation – Figure 1.2) controlled simultaneously. This complexity makes it necessary to investigate dragonfly aerodynamics using live specimens to obtain a deeper understanding of dragon-fly aerodynamics for aircraft design (Chin & Lentink 2016; Hedrick et al. 2015).

A few groups have attempted full-span qualitative and quantitative stud-ies involving live dragonflies (Bomphrey et al. 2016; Reavis & Luttges 1988; Somps & Luttges 1985, 1986; Thomas et al. 2004; Yates 1986). Thomas et al. (2004) studied the aerodynamics of tethered and free-flying dragonflies qualitatively using smoke visualization in a wind tunnel whereas Bomphrey et al. (2016) did so using particle image velocimetry (PIV).

Thomas et al. (2004) reported that the forewing LEV spans from wingtip to wingtip, while the hindwing typically exhibits attached flow. However, Lai and Shen (2012) hypothesized that the LEV extending over the body could be caused by the wind tunnel flow and the tilted body alignment of the dragonfly during take-off. Hence, Hefler et al. (2018) studied dragonflies in vivo in a still air flight chamber to decouple the possible influence of headwind.

In this section, we present the wing-wing aerodynamic interactions of a dragonfly *Pantala flavescens* in a spanwise resolved manner that helps highlight specific features of this high-AR, root-flapping wing system (Hefler et al. 2020; Hefler et al. 2018). Previous studies found that when the hindwing is leading the phasing by approximately 90° (in slow to moderate forward flight or while hovering), the interaction between the forewing and the hindwing significantly affects only the hindwing (Broering et al. 2012; Hsieh et al. 2010; Hu & Deng 2014; Maybury & Lehmann 2004; Rival et al. 2011; Xie & Huang 2015; Zheng et al. 2016a, 2016b). Therefore, regarding wing-wing interactions in normal slow forward flight and hovering, we evaluate how the presence of the forewing affects the hindwing.

4.1 Flow Features during Slow Forward Flight and Hovering

During slow forward flight and hovering, dragonflies adopt an approximately quarter cycle hindwing lead phasing and a horizontal or slightly inclined body posture (Azuma & Watanabe 1988; Norberg 1975; Wakeling & Ellington 1997). Hefler et al. (2018) also observed that in this flight mode, the hindwing typically flaps more to the ventral side than the forewing.

Particle image velocimetry flow measurements on tethered dragonflies found distinct interaction features in the inner and the outer span for the typical case of quarter-cycle hindwing-led phasing (Hefler et al. 2018). These regions along the span cannot be clearly isolated as transition regions border them. Two transitions are reported in the root region and in the mid-span region, respectively. Figure 4.2 shows the four regions and a 3D schematic of the flow structures as reported in Hefler et al. (2018): the root region, the inner region, the transition region and the outer region. Lastly, a recent numerical study reported span-resolved aerodynamic forces related to the varied interaction features along the wingspan (Hefler et al. 2020).

A distinct LEV-LEV interaction has been identified in most parts of the inner half of the hindwing span of the dragonfly (Hefler et al. 2018). Differently, on the outer span, the hindwing captures the forewing-shed TEV (Hefler et al. 2018). Figure 4.3 presents the process of the vortex

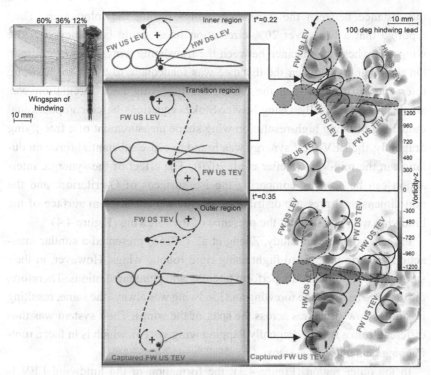

Figure 4.2 Particle image velocimetry flow measurement of cross sections along the span of a dragonfly. On the left side is the definition of the hindwing span (S_{HW}), and the root (red), inner (yellow), transition (green) and outer (purple) regions. On the right is a schematic of the characteristic interactions and the 3D flow structures over the span. FW, HW, US and DS denote forewing, hindwing, upstroke and downstroke, respectively. Adapted from Hefler et al. (2018) with permission of the *Journal of Experimental Biology*.

interaction measured in the inner span at 18% from the wing root during downstroke and the outer span vortex capture at 78% span in a similar manner. The LEV-LEV interaction is facilitated by the dynamic cambering of the forewing as well as the short distance between the wings in the translational phase of the flapping cycle. The dynamic cambering during the stroke reversal of the forewing allows the forewing LEV to be sustained and shed along the pressure side (e.g. the bottom surface during downstroke) of the forewing well after the stroke reversal. Due to the hindwing-led phasing, this shedding forewing LEV moves next to the LEV of the hindwing, making energy transfer between the shear layers possible. By measuring the circulation of the hindwing LEV with and without the effect of the forewing, Hefler et al. (2018) established that the modulation of the hindwing LEV circulation depends on the distance between the translating wings during the interaction. They reported that a little over 20% circulation boost could be achieved at the wingspan where the distance between the translating wings is most optimal. On the other hand, when the distance was inadequate to allow for a strong interaction effect between the LEVs, the hindwing LEV circulation was attenuated substantially. Using Navier-Stokes equations-based computational modelling utilizing high-resolution wing shape measurement of a free-flying dragonfly, the LEV-LEV synergy was found to cause substantial force modulation in the mid-span (Hefler et al. 2020). The effect of the synergic interaction can be seen by comparing the iso surfaces of Q-criterion, and the non-dimensional pressure distribution on the top and bottom surface of the hindwing with and without the presence of the forewing (Figure 4.4).

In their experimental study, Zheng et al. (2016b) presented a similar interaction scenario in forward flight using rigid robotic wings. However, in their study, the wings could only perform pitching and plunging motions. Therefore, the distance between the forewing and hindwing was always the same, resulting in similar flow structures across the span of the wings. Their system was thus different from a natural dragonfly flapping wing system, which is in fact a root-flapping tandem-wing system.

In the outer region (Figure 4.3), the formation of the hindwing LEV is delayed or even suppressed by the downwash effect of the forewing. The hindwing, however, is able to form an LEV in the last quarter of the stroke via vortex capture. This vortex capture is essentially different from wake capture, in which the wing utilizes the flow of its own vortical structures shed during stroke reversal (Birch & Dickinson 2003; Dickinson et al. 1999; Lehmann et al. 2005). Vortex capture could provide a useful aerodynamic force and a momentum that eases the pitching motion of the wing (Lehmann 2008). The capture of the forewing TEV was also found in numerical simulation (shown in Figure 4.5) of a free-flying dragonfly (Hefler et al. 2020). Further to the capture mechanism, the 3D flow field of the numerical model revealed that the positioning of the 'tail' of the shed TEV of the forewing and the hindwing is such that it facilitates the formation of a narrow jet in the wake, resulting in a particularly large thrust gain by the tandem winged system (Hefler et al. 2020).

Hsieh et al. (2010) also reported vortex capture while the wings were counter-stroking. However, in their 2D numerical study, a TEV-LEV pair formed simultaneously by the forewing from which the TEV was captured by the hindwing. Because of this interaction, the hindwing lift was boosted to 109% of the single-wing maxima for a short period of time at the end of the upstroke (Hsieh et al. 2010). Zheng et al. (2016b) also showed that the

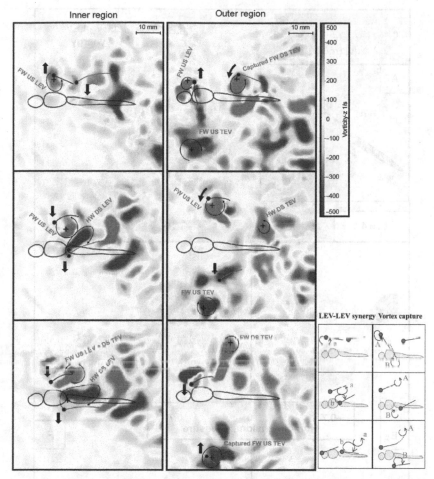

Figure 4.3 Vortex interactions in the inner span and the outer span measured by particle image velocimetry. FW, HW, US and DS denote forewing, hindwing, upstroke and downstroke respectively. Black arrows indicate the wing motion. A schematic diagram shows the interaction mechanisms for an easier interpretation of the presented flow fields. Adapted from Hefler et al. (2018) with permission of the *Journal of Experimental Biology*.

counter-stroking hindwing can capture and utilize the TEV of the forewing during forward flight. In their study, the forewing TEV was captured about halfway through the stroke. They found that the vortex capture generated a force that was as high as that from the LEV-LEV interaction (about 39%) in the case of 90° wing phasing (Zheng et al. 2016b). This suggests that an LEV-LEV vortex synergy and the vortex capture could be equally important tools for dragonflies that can result in comparable force gains. This observation is

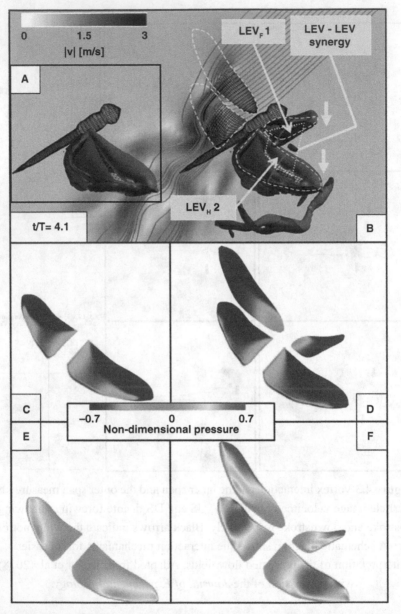

Figure 4.4 The LEV-LEV synergy and its effect on the hindwing by comparing flows and surface pressures of a free-flying dragonfly in solo hindwing flight and tandem winged flight. The numerical results are based on the high-resolution direct linear transformation technique kinematics and wing morphological measurement of a free-flying dragonfly. (a) Iso surfaces of Q-criterion (Q = 1.2 coloured by Y-vorticity to indicate direction; red is positive and blue is negative). (b) Iso surfaces of Q = 1.2, streamlines and non-

also supported by recent numerical investigation of Hefler et al. (2020). Xie and Huang (2015) also observed vortex capture at 135 and 270° phasing. Different from the current findings, however, the vortex shed by the forewing interacted with the hindwing after its stroke reversal and induced the formation of a new LEV during the next stroke. Xie and Huang (2015) used 2D modelling and linked the interaction scenarios to the effect of wing phasing but not to the different regions of root-flapping wings.

By measuring the flow field relatively close to the wing root, Hefler et al. (2018) also found that in the root region, the LEV-LEV interaction was not yet present. The wings essentially operate as one slotted wing despite flapping out of phase. The LE of the forewing guided and distributed the incoming flow while forming a weak LEV (Figure 4.6). The hindwing, however, did not generate an LEV. A TEV was created by the pitching motion of the hindwing. The upstroke and downstroke counter-rotating hindwing TEVs formed a narrow, reverse von Kármán vortex street that induced a streamwise jet. Moving away from the root, the phased operation of the wings increases the size of the gap between the wings, allowing the hindwing LEV to form and the LEV-LEV interaction to take place. Therefore, the region closest to the root can be considered as a transition from a slotted single wing-like set-up to the previously described inner span region dominated by the LEV-LEV interaction.

Another transition was found around the mid-span between the region of the LEV-LEV interaction and the region where the vortex capture takes place (Hefler et al. 2018). Here, despite having no direct interaction with the shed vortex, the hindwing moves through the flow induced by the forewing, which has a substantial streamwise component (Figure 4.6). This streamwise flow reduces the effective AoA of the hindwing and suppresses LEV formation (Hu & Deng 2014; Maybury & Lehmann 2004; Sun & Lan 2004). These aforementioned works did not connect the forewing downwash effect to any particular region of the wings.

Caption for Figure 4.4 (cont.)

dimensional absolute velocity; and pressure distribution on the top (C–D) and bottom (E–F) wing surface close to the end of the downstroke of the hindwing in solo (A–C–E) and tandem mode (B–D–F). The wings' motion is indicated by yellow arrows, and the mid-span is highlighted by the pink line for reference. t/T is the dimensionless time, where T is the period and t = 0 at the start of the downstroke of the forewing in the first flapping cycle. IvI is the velocity magnitude at the mid-span cross section. Reprinted from Hefler et al. (2020), with the permission of AIP Publishing.

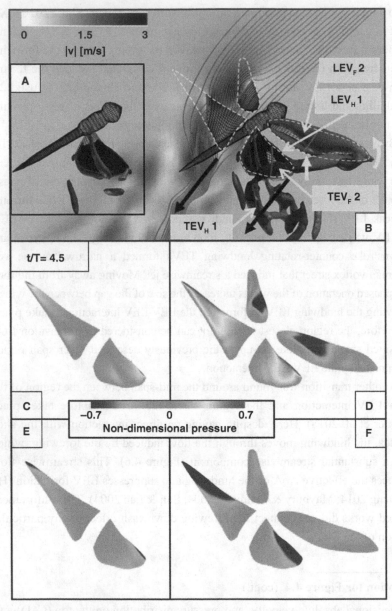

Figure 4.5 The combined effect of the previous LEV-LEV synergy and vortex capture on the hindwing by comparing flows and surface pressures of a free-flying dragonfly in solo hindwing flight and tandem winged flight. The numerical results are based on high-resolution direct linear transformation technique kinematics and wing morphological measurement of a free-flying dragonfly. (a) Iso surfaces of Q-criterion (Q = 1.2 coloured by Y-vorticity to indicate direction; red is positive and blue is negative). (b) Iso surfaces of

4.2 Implications of Wing Flexibility in Facilitating Wing-Wing Interactions

Experimental and numerical investigations using rigid wings in tandem laid down some of the fundamental knowledge about inter-wing aerodynamic effects. It was shown that a rigid wing can induce a strong TEV in its wake. The TEV of the forewing can interact with the LEV of the hindwing (Broering et al. 2012; Hsieh et al. 2010; Hu & Deng 2014; Maybury & Lehmann 2004; Rival et al. 2011; Xie & Huang 2015; Zheng et al. 2016b). This kind of TEV-LEV interaction can strengthen the hindwing LEV. On the other hand, a dragonfly wing is flexible, and its shape is the result of the dynamic balance between the wing's inertia, stiffness and the aerodynamic force exerted on it (Combes & Daniel 2003a, 2003b; Kang et al. 2011). A compliant wing adapts its shape to the flow field and does not shed as much vorticity from the TE as a rigid wing does (Fu et al. 2018; Heathcote & Gursul 2007). Hence, a flexible wing has less TE-induced vorticity, allowing sufficient space for the forewing LEV and the hindwing LEV to interact.

In natural dragonflies, the LEV of each wing starts to shed along the wing surface while the wing is pitching at the end of a stroke; it then moves towards the TE on the wing's pressure side and initiates the TEV formation of the next stroke (see Figure 4.3). The LEVs after stroke reversal would be quickly broken down by the translational motion of a rigid wing (Maybury & Lehmann 2004), while a new LEV-TEV pair would be formed by the next stroke. In contrast, a compliant wing membrane would envelop the shedding LEV (Figure 4.3) on the pressure side of the wing, conserving it. Free-flight measurements of dragonflies in a wind tunnel also confirm that rather than forming a distinct starting vortex simultaneously with the LEV, a sheer layer forms behind the TE that rolls up, under Kelvin-Helmholtz instability, into a series of transverse vortices with circulation of opposite sign to the circulation around the wing and

---

Caption for Figure 4.5 (cont.)

Q = 1.2, streamlines and non-dimensional absolute velocity; and pressure distribution on the top (C–D) and bottom (E–F) wing surface during the upstroke of the hindwing in solo (A–C–E) and tandem mode (B–D–F). The wings' motion is indicated by yellow arrows, and the mid-span is highlighted by the pink line for reference. t/T is the dimensionless time, where T is the period and t = 0 at the start of the downstroke of the forewing in the first flapping cycle. IvI is the velocity magnitude at the mid-span cross section. Reprinted from Hefler et al. (2020), with the permission of AIP Publishing.

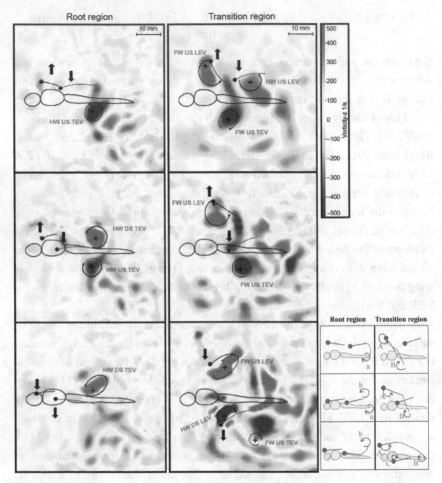

Figure 4.6 Transient flow features in the root region and the transition region measured by particle image velocimetry. FW, HW, US and DS denote forewing, hindwing, upstroke and downstroke, respectively. Black arrows indicate the wing motion. A schematic diagram shows the interaction mechanisms for an easier interpretation of the presented flow fields. Adapted from Hefler et al. (2018) with permission of the *Journal of Experimental Biology*.

LEV (Thomas et al. 2004). The lack of presence or delayed formation of the TEV leaves space for the forewing LEV and the hindwing LEV to interact in dragonflies to promote hindwing LEV circulation.

4.3 Wing-Wing Interactions during Free Flight

It is important to assess how representative the flow features observed in tethered flight experiments are of those displayed during unrestricted flight. Hefler et al.

(2018) conducted free-flight measurements on the same species of live dragonflies to support their findings. Ensuring that measurements are taken from precise locations of the wing is difficult when working with free-flying specimens. Nevertheless, the characteristic regions that were identified using tethered experiments are wide enough for an observer to determine to which region a particular free-flight measurement belongs. Free-flight measurements did confirm that the LEV of the forewing interacts with the hindwing in a way that it feeds the sheer layer of the LE of the hindwing (Hefler et al. 2018). Regarding the inner span during slow forward flight and hovering, this observation somewhat contradicts the previous findings by Hu and Deng (2014) and Maybury and Lehmann (2004) that the TEV of the forewing would be the major player in the inter-wing direct vortex interactions. It is possible, however, that under different flight conditions when headwind exists, or during take-off and extreme mid-air manoeuvring, the forewing could develop a strong TEV that consequently interacts with the hindwing primarily instead of the forewing LEV.

Thomas et al. (2004) and Bomphrey et al. (2016) recorded the flight of dragonflies in a wind tunnel during steady flight and manoeuvring. The commonly observed flow dynamics of 'normal free flight' were characterized by an LEV on the forewing during downstroke, attached flow on the forewing during upstroke and attached flow on the hindwing throughout (Figure 4.7). Additionally, Thomas et al. (2004) reported a number of other aerodynamic patterns such as the formation of a hindwing LEV, attached flow on the forewing and even stalled flow that points out the complexity that exists in tandem winged insect flight in an environment where external flow exists.

Regarding the outer span, free-flight flow field measurements suggest that the successful capture of a shed forewing vortex is dependent on the flight direction (Hefler et al. 2018). For example, in descending flight shown in Figure 4.8, the vortex that is shed by the forewing above the dragonfly (FW DS TEV) moves away from the direction of flight so that the hindwing is less able to capture and utilize it during its upstroke. On the other hand, the downstroking hindwing is more likely to capture the vortex that is shed by the forewing below the dragonfly (FW US TEV) as it flies towards it (Figure 4.8).

4.4 Summary and Concluding Remarks

Several works have aimed to characterize insect-scale tandem wing flapping flight. Experimental and numerical studies have established a strong background for numerous possible interaction scenarios according to the wide range of possible kinematic and geometrical set-ups as well as external

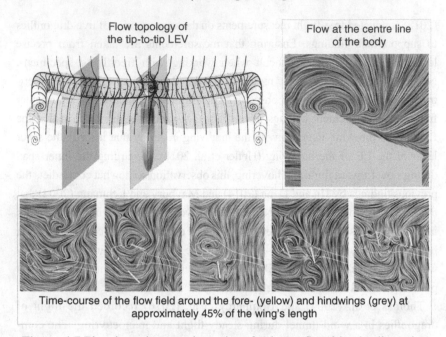

Figure 4.7 Flapping wing aerodynamics of a dragonfly with a leading-edge vortex over the forewings and thorax and attached flow over the hindwings. (a) Topology of the leading-edge vortex of dragonfly as described by Thomas et al. (2004). (b) Cross section of the flow at the centre line of the body measured by particle image velocimetry, with instantaneous streamlines visualized by line integral convolution. (c) Time course of the measured flow field around the forewings (yellow) and hindwings (grey) with the laser sheet incident at approximately 45% of the wing's length from hinge to tip. The leading-edge vortex on the forewing is observed clearly, while the flow remains attached to the hindwing. Adapted from Bomphrey et al. (2016).

environmental conditions. Modulation of the generated aerodynamic forces by these interaction scenarios can be substantial(Hefler et al. 2020; Hu & Deng 2014; Maybury & Lehmann 2004).

Investigating 3D root-flapping flexible wing set-ups that incorporate the conscious adjustments of flapping kinematics seems impossible. Using live specimens has inherent difficulties as well. Nevertheless, dragonflies, as the most agile and versatile tandem winged species, have been successfully studied experimentally in recent years (Bomphrey et al. 2016; Chen et al. 2013; Hefler et al. 2020; Hefler et al. 2018; Huang & Sun 2007; Kumar et al. 2019; Thomas et al. 2004). These works present interesting new findings and aspects.

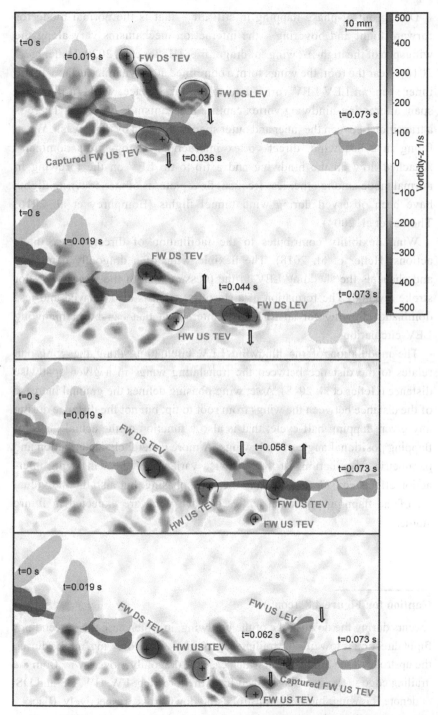

Figure 4.8 Free-flight flow features measured by particle image velocimetry in the outer span of a descending dragonfly. The outer span vortex capture only

During out-of-phase flapping in still air – that is the normal mode for forward flight and hovering – the interaction mechanisms vary along the wingspan of the high-AR wings of dragonflies (Hefler et al. 2020; Hefler et al. 2018). Near the root, the wings form a combined aerodynamic surface. In the inner span, an LEV-LEV vortex interaction takes place, while in the outer span, a forewing-hindwing vortex capture mechanism is observed. There is a transition between the inner and outer span dominated by downwash. When flying in a headwind, direct vortex interactions become less dominant. Attached flow on the hindwing and a tip-to-tip LEV on the forewing in normal mode among other less frequent flow patterns of manoeuvring flights have been observed during wind tunnel flights (Bomphrey et al. 2016; Thomas et al. 2004).

Wing flexibility contributes to the facilitation of direct vortex interactions (Hefler et al. 2018). The flexible wing of a dragonfly conserves and channels the shedding LEV on the pressure side of the wing well after stroke reversal. The formation of a starting vortex is also delayed and less dominant, allowing for an LEV-LEV interaction that can boost hindwing LEV circulation.

The modulation of the hindwing LEV circulation along the wingspan relates to the distance between the translating wings in a given spanwise distance (Hefler et al. 2018). A set wing phasing defines the gradual increase of the distance between the wings from root to tip, but not the distance during any given flapping half-cycle; that is also a function of the actual range of flapping positional angle of both wings. A more general characterization and parametric examinations of the spanwise variance of the wing-wing interaction effects considering not only the wing phasing, but also the individual set of the flapping positional angles of each wing are expected in future studies.

Caption for Figure 4.8 (cont.)

occurs during the downstroke of the hindwing, in which case the descending flight direction is towards the trailing edge vortex of the forewing, unlike during the upstroke of the hindwing, in which case the dragonfly moves away from the trailing edge vortex previously shed from the forewing. FW, HW, US and DS denote forewing, hindwing, upstroke and downstroke, respectively. Black arrows indicate the wing motion. Adapted from Hefler et al. (2018) with permission of the *Journal of Experimental Biology*.

5 Effects of Aspect Ratio on Flapping Wing Aerodynamics

The unsteady 3D fluid physics associated with flapping wing aerodynamics have been probed in a large number of previous studies (Chin & Lentink 2016; Platzer et al. 2008; Sane 2003; Shyy et al. 2013). Several unsteady 3D mechanisms that are responsible for enhancing the lift of a flapping wing are notably delayed stall of an LEV (Ellington et al. 1996) and recapturing of its own wake (Dickinson et al. 1999), as discussed in Section 1. Both phenomena depend strongly on the wing kinematics as well as the flow parameters such as the Reynolds number. For example, the LEV can be stabilized by spanwise flow in its core at a high Reynolds number and by induced flow by the TiV at a low Reynolds number (Shyy et al. 2013).

In fixed-wing aerodynamics, the *AR* is one of the key parameters characterizing the 3D flow mechanisms. It is established that more force per area and higher efficiency can be obtained with high-*AR* wings while low-*AR* wings suffer from induced drag due to TiVs. On the other hand, many biological flyers tend to have lower-*AR* wings operating at high AoAs in the low-Reynolds-number regime, still capable of generating aerodynamic forces efficiently (Shyy et al. 2013). With such flying characteristics, the low-Reynolds-number unsteady wing motions create interesting phenomena deserving of in-depth investigation.

Studies of *AR* effects on rotating/flapping wings undergoing unsteady motions have been conducted both experimentally (Carr et al. 2013; Carr et al. 2015; Fu et al. 2014; Han et al. 2015; Kruyt et al. 2014; Kruyt et al. 2015; Ozen & Rockwell 2013; Phillips et al. 2015; Usherwood & Ellington 2002) and numerically (Garmann & Visbal 2014; Harbig et al. 2014; Luo & Sun 2005). A continuous vortex loop formed by a stably attached LEV, a TiV and a TEV is observed (Carr et al. 2013; Garmann & Visbal 2014; Harbig et al. 2014). The LEV gradually lifts off the wing surface and shifts aft as the *AR* increases, resulting in LEV break-down and detachment at the outboard region of slender wings (Carr et al. 2013; Fu et al. 2014; Phillips et al. 2015). Fu et al. (2014) studied the LEV structures, stability and circulation for rotating wings with varied *AR*s. The LEV structures were likely to be dominated by the chord-normalized dimensionless distance from the rotation centre. Kruyt et al. (2015) showed that the LEV remains attached at a high AoA within four chord lengths of the local radius and separates further outboard on wings with *AR* higher than 4. On the other hand, the slender wing simultaneously brings on a decreased 3D effect and an increased LEV breakdown, leading to a modest influence of *AR* on the lift and drag forces (Luo & Sun 2005; Usherwood & Ellington 2002). Overall, the *AR*'s role in the aerodynamic performance of a flapping wing is to be clarified.

Most studies in low-*AR* flapping wing aerodynamics have focused specific-ally on aerodynamic force generation due to the LEV and TEV. In this section, we discuss additional physical mechanisms associated with *AR* effects in the operation of flapping wings. Firstly, we consider the interplay between the wing kinematics, TiV and aerodynamic force generation at a fixed *AR* (Shyy et al. 2009). Then, we consider the role of the *AR* in the resulting vortex dynamics.

5.1 Tip-Leading-Edge Vortex Dynamics

Tip vortices associated with fixed finite wings are traditionally seen as phenom-ena that decrease lift and induce drag (Anderson 2011). For low-*AR* flapping wings, Shyy et al. (2009) discovered that TiVs can increase lift both by creating a low-pressure region near the wingtip and by anchoring the LEV to delay or even prevent it from shedding when a large-amplitude pitching motion is delayed. For small-amplitude, synchronized pitching motions, the LEV remains attached along the spanwise direction and the tip effects are not prominent; in such situations, the aerodynamics are little affected by the *AR* of a wing.

Shyy et al. (2009) considered a rigid wing with a prescribed plunge motion with an amplitude normalized by the chord h_a/c and an active pitching motion with an amplitude α_a and a phase lag φ between the pitching and plunging motions. The role and implications of changing kinematic parameters on the aerodynamics of such a wing and the associated 3D fluid physics were illus-trated with two representative cases: (i) a large-amplitude, delayed rotation which sees that the pitching motion lags that of the translation (plunging) motion (2 h_a/c = 2.0, α_a = 45° and φ = 60°); and (ii) a small-amplitude, synchronized rotation where the pitching and translation are in phase (2 h_a/c = 3.0, α_a = 80° and φ = 90°). The *AR* based on the distance between the wing root and tip is 2, and the Reynolds number of 100 is based on the chord and the maximum plunge velocity. Shyy et al. (2009) observed that the flow structures and the aerodynamics are significantly different between these two cases, largely due to the impact of the TiVs. In contrast to the delayed rotation, the synchronized rotation with low AoAs can largely negate the effect of the TiVs.

Figure 5.1 compares the vorticity contours of the 2D case with the 3D simulations at mid-span and the tip at the end of the stroke for a delayed rotation case. Unlike the 3D simulation, the 2D case eliminates any possible TiV effects. Two kinematic patterns are shown; the left case depicts qualitatively different time histories in lift between an infinite (2D) and a finite (3D) wing, while the right case demonstrates that the lift history of an infinite wing can closely mimic that of a low-*AR* finite wing. The difference in the flow physics encountered due to 3D phenomena is noticeable. Through the entire stroke, it is seen that the shed vortices are more dissipative in 3D. Shyy et al. (2009) also observed the

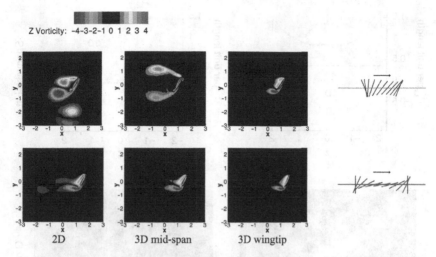

Figure 5.1 Z-vorticity contours at the end of the stroke of the (top) delayed rotation case and (bottom) synchronized rotation case. Results of a comparative study of 2D and 3D computational fluid dynamics. Adapted from Shyy et al. (2009).

behaviour of the vortices changes, as evidenced by the snapshots at different time instances in the stroke. In 2D, the stroke starts by running into the previously shed vortices, whereas in 3D, the rotational starting vortex (RSV) is shed above the plane of translation and convected away from one another due to the influence of the tip vortices. The variation along the spanwise direction is weak, and the differences between the 2D and 3D simulations are small.

A perspective of the 3D spanwise variation is given in Figure 5.2, where the second invariant of the velocity gradient tensor, the Q criterion (Shyy et al. 2009), is shown to illustrate the vortical nature of the flow. The spanwise variation in the 3D case shows remarkable changes in the aerodynamic loadings. The RSVs stay anchored at the tips. The non-dimensional lift per unit span due to pressure versus the 2D equivalent suggests that the TiVs can enhance the lift for the majority of the stroke cycle. For the synchronized rotation case, the lift response is almost uniform across the rest of the wing. The flow features do not feature much variation in the spanwise direction compared to the delayed rotation case. The high angular amplitudes lead to low AoAs and, coupled with the timing of the rotation, lead to a flow that not only lacks a dominant response due to the TiVs, but also does not experience delayed stall as the formation of the LEV is not promoted. The timing of the rotation for this example puts the flat plate at its minimum AoA at maximum translational velocity, while the translational velocity is zero when the flat plate is vertical.

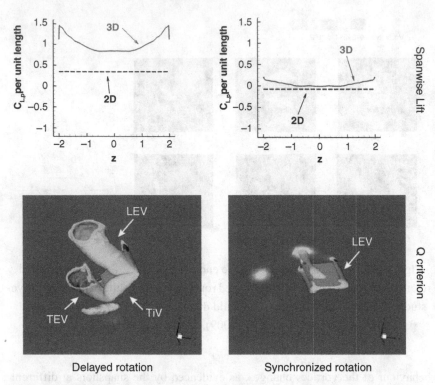

Figure 5.2 The lift per unit span and iso- surfaces (= 0.75) snapshots of (left) the delayed rotation and (right) synchronized rotation cases at the mid-stroke. The spanwise variation in forces is compared with the 2D equivalent marked for reference. Computational fluid dynamics results. Adapted from Shyy et al. (2009).

The instantaneous lift coefficient for the two cases examined is illustrated in Figure 5.3. In the first case – that is delayed rotation – Shyy et al. (2009) suggested that the TiVs played a dominant role in the aerodynamic loading, and in particular the lift was enhanced significantly near the wingtips because of the presence of strong TiVs as well as their secondary influence of anchoring the RSVs. Compared to an infinite wing, the TiVs caused added mass flux across the span of a low-*AR* wing, which helps push the shed RSV and TEV at mid-span away from one another. Furthermore, a spanwise variation occurs in the effective AoA induced by the downwash, stronger near the tip. Overall, the TiVs allowed the RSVs in their neighbourhood to be anchored near the wing surface, which promotes a low-pressure region and enhances lift. On the other hand, the aerodynamic loading of a low-*AR* wing is well approximated by the 2D calculations for the synchronized rotation case.

Unlike traditional fixed-wing aircraft, the TiVs of a low-*AR* flapping wing can either promote or make little impact on the aerodynamics. The timing of the

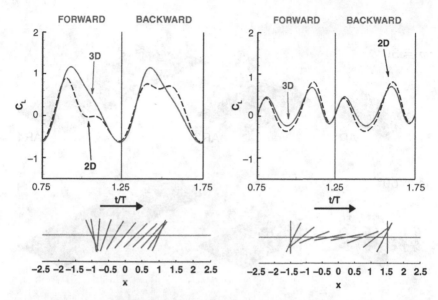

Figure 5.3 Instantaneous 2D and 3D lift histories for (left) delayed rotation and (right) synchronized rotation. Here, T denotes a flapping period. The time instance $t/T = 0.75$ corresponds to the start of the forward stroke and $t/T - 1.25$. Adapted from Shyy et al. (2009).

rotational motion of the wings in relation to the translational motion affects how strong the 3D TiV effects are.

5.2 Vortex Evolution for Low- and Intermediate-AR Wings

The low-AR wing vortex dynamics, including the LEV, the TiV and the interaction between them, was further studied by Fu et al. (2017) using PIV measurements on rotating wings in still water. The wingtip-based Reynolds number was kept at $Re = 5.3 \times 10^3$ and the AoA at $\alpha = 45°$. The wing AR was varied between 1 and 4.

The vortical features of the 3D unsteady flow around the rotating wings are illustrated in Figure 5.4. For each wing planform, an LEV forms and is attached to the leeward side near the LE, while a TEV gradually detaches from the wing. The size of the LEV increases along the span up to the wingtip. The LEVs encounter significant growth as the wing rotates from $\varphi = 30°$ to $\varphi = 60°$ and remain stably attached in the leeward side near the LE with coherent vortex structures maintained with the exception of the case of $\varphi = 60°$ for $AR = 2$ and 4. The size of the LEV grows as the AR increases. The shrinking of the LEV near the wingtip is due to the presence of the TiV, and the underlying mechanism is described further in what follows.

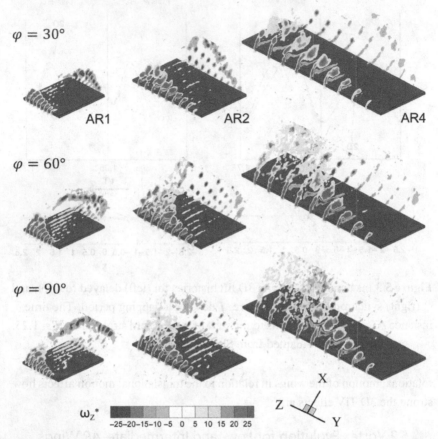

Figure 5.4 Slices of normalized spanwise vorticity fields against the span for $AR = 1$, 2, and 4 flat plates at azimuthal angles $\varphi = 30°$, 60° and 90°. Particle image velocimetry experimental results. Adapted from Fu et al. (2017) Reprinted by permission of the American Institute of Aeronautics and Astronautics, Inc. © AIAA.

In previous studies on the effects of AR (Carr et al. 2015; Fu et al. 2014; Phillips et al. 2015), the LEV structure was described with the dimensionless length $z_b^* = z/b$, with the wingspan b as the reference length to normalize the local radius z. However, Fu et al. (2017) show that the flow typology became similar across the considered ARs between 1 and 4 when the flow features were described with $z_c^* = z/c = AR\, z_b$, where c is the chord, as shown in Figure 5.5, suggesting the existence of a key scaling parameter that includes the effects of the AR. In the inner span region, the centre of the LEV moves along a dimensionless trajectory when normalized by the wing chord. In the outer span region, the trajectory is influenced by the TiV and in turn by the AR. For a wing having a larger AR, the region affected by the TiV is smaller in relation to

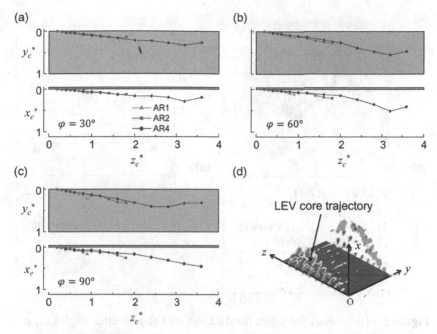

Figure 5.5 Leading-edge vortex centre trajectories in the wing-based coordinate at azimuthal angles at (a) $\varphi = 30°$, (b) $\varphi = 60°$ and (c) $\varphi = 90°$. The wing-based coordinate is illustrated in (d). Adapted from Fu et al. (2017). Reprinted by permission of the American Institute of Aeronautics and Astronautics, Inc. © AIAA.

the complete wing area. As already mentioned, and in contrast to a fixed wing, the TiV of a rotating (flapping) wing has a benign effect by anchoring the LEV to the wing surface in the inboard region via its downwash. At the late stage, the LEV breakdown on the high-AR wing yields the breakdown of the TiV, followed by lift-off of the LEV in the inboard region.

The relation between the vortex dynamics and the velocity field was discussed by Fu et al. (2017) by considering the normalized downwash velocities (Figure 5.6). Whereas the downwash shows similar magnitudes for the wing with $AR = 2$ and $AR = 4$ at azimuthal angles $\varphi = 30°$ and 60°, the wing with $AR = 4$ shows a noticeably smaller downwash magnitude than that of the wing with $AR = 2$ at $\varphi = 90°$. The downwash is associated with both the LEV and the TiV. Thus, it is difficult to tell apart their contributions. Fu et al. (2017) argue that the velocity deviation is attributed to the loss of the downwash induced by the TiV when $AR = 4$, based on the finding by Fu et al. (2014) that the LEV circulation in the inner half span experiences little change as the wing rotates from $\varphi = 60°$ to $\varphi = 90°$ under the same kinematics. During this deceleration phase, the outboard LEV breaks down on an $AR = 4$ wing, which causes the breakdown of the TiV.

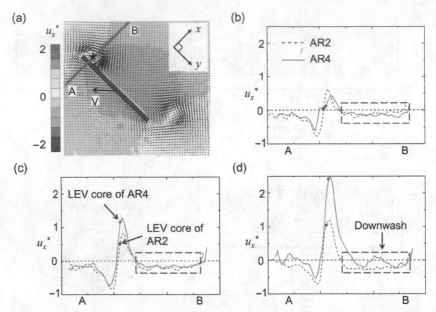

Figure 5.6 Downwash velocities for $AR2$ and $AR4$ at azimuthal angles (b) $\varphi =$ 30°, (c) $\varphi = 60°$ and (d) $\varphi = 90°$ at $z_c^* = 0.8$. The downwash refers to negative velocity in the x-direction. The u_x transect AB in (b), (c) and (d) are indicated in (a) and highlighted in red. The black star represents the location of the leading-edge vortex centre. Particle image velocimetry experimental results. Adapted from Fu et al. (2017). Reprinted by permission of the American Institute of Aeronautics and Astronautics, Inc. © AIAA.

Past TiV breakdown, the spatial similarity with respect to the AR in the x-direction breaks while in the y-direction the spatial similarity is maintained. This is because the downwash induced by the TiV vanishes with breakdown of the TiV, causing lift-off of the LEV on the wing with $AR = 4$.

The fluid dynamic forces acting on the wings are shown in Figure 5.7. All planforms show a lift plateau at the early acceleration stage of the period due to the added-mass effect. Lower-AR wings experience a larger added-mass effect. However, a wing with a higher AR shows a larger lift coefficient afterwards, especially prior to the last quarter period when the LEV closely attaches on the wing. The average lift coefficient and the lift-to-drag ratio (Figure 5.7b) also increase with AR in the considered AR range, verifying the significant lift enhancement effect of the LEV and that the wing with higher AR benefits from more vortex lift due to the larger vortex structure in the outboard region. During the deceleration phase, the lift coefficient of $AR4$ rapidly decreases and becomes similar with that of $AR2$ afterwards, correlating to the LEV

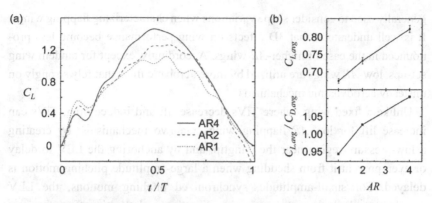

Figure 5.7 Fluid dynamic forces acted on the wing. (a) Time courses of the lift coefficient during one period. (b) The change of the averaged lift coefficient and lift-to-drag ratio with AR. Adapted from Fu et al. (2017). Reprinted by permission of the American Institute of Aeronautics and Astronautics, Inc. © AIAA.

detachment in the outboard region. The outboard detachment of the LEV results in a stall encountered in the outer span on $AR4$.

Fu et al. (2017) note that efficiency reduces when the AR is further increased, because of the stall encountered in the outer span. These competing effects suggest the existence of an optimal AR which provides both high lift coefficient and high efficiency. The LEV breakdown and detachment lead to upward migration of the local LEV centre towards the LE. Comparison of the LEV centres at $\varphi = 60°$ and $\varphi = 90°$ in Figure 5.5 reveals that the upward migration occurs beyond $z_c^* = 2.8$, suggesting LEV detachment in this region. This LEV detachment also can be seen by the vorticity field shown in Figure 5.4. Since the LEV detachment occurs at around $z_c^* = 2.8$, Fu et al. (2017) conclude that this value can also offer a good estimate of the optimal AR under the investigated conditions. The estimated optimal AR matches with previous findings by Kruyt et al. (2015), Han et al. (2015) and Lee et al. (2016). This value also matches the wing AR of some insect species such as fruit flies, honeybees and hawkmoths (Shyy et al. 2010).

5.3 Summary and Concluding Remarks

Nature showcases an almost endless variety of wing sizes and forms. However, through evolution aerodynamic performance was not the only consideration. Wings may play an important role in mating (Alexander & Brown 1963) and provide camouflage or shielding (Combes 2010; Wootton 1992). The AR is generally used to consider shape variations when characterizing flapping wings. It is well understood that 3D effects on wing performance become less pronounced in the case of higher-*AR* wings. Accordingly, except for tandem wing set-ups, low-*AR* wings are utilized by more acrobatic fliers that rely strongly on unsteady aerodynamic mechanisms.

Unlike a fixed wing, where TiVs decrease lift and induce drag, TiVs can increase lift for low-*AR* flapping wings via two mechanisms: by creating a low-pressure region near the wingtip, and by anchoring the LEV to delay or even prevent it from shedding when a large-amplitude pitching motion is delayed. For small-amplitude, synchronized pitching motions, the LEV remains attached along the spanwise direction and the tip effects are not prominent; in such situations, the aerodynamics are little affected by the *AR* of a wing.

For a rotating 3D wing, the size of the LEV monotonically increases along the spanwise direction; it also grows as the *AR* increases. Near the wingtip, the LEV interacts with the TiV and the interaction shrinks its size.

Despite our already rich understanding of *AR* effects in flapping flight, most studies considered a rather narrow range of wing *AR*. With advancements in material science, it could soon be feasible to consider higher-*AR* flapping wing set-ups for MAVs without risking structural integrity. Furthermore, considering tandem wing set-ups, *AR* effects on wing-wing interactions are still to be researched.

6 Perspectives and Future Outlook

In the previous sections, we have discussed four particular topics related to insect-scale flight focused on recent advancements in our theoretical understanding and characterization. Thanks to continuous support and interest within the academic and industrial research communities, we have been witnessing an accelerated progress in overall understanding and practical implementation of the subject matter.

While insect-scale flyers experience highly varied flight conditions, their behaviours in aerodynamics as well as flight planning, sensing and control can be cast in a quasi-steady framework, taking the environmental fluctuations as time-dependent boundary conditions. However, higher-order averaging techniques may be needed to understand the dynamics of flapping wing flight and to analyse its stability. Although progress has been made to address the intrinsic interactions between flexible wing structures, flight stability and associated aerodynamics, more exploration is needed. In particular, with the rapid growth of data, machine learning techniques and non-physical, model-based control techniques can add significant value to our understanding and capabilities (Shyy et al. 2016).

As presented, a low-AR flapping wing also exhibits noticeably different behaviours compared to a fixed wing, including flow structures, much enhanced lift due to coherent vortical flow structures and related pressure effects, and spanwise variations including the possibly benign effect of TiVs. The coordination between plunging, pitching and sweeping parameters is critical for aerodynamic performance of flapping wings. As highlighted, for a rigid wing, advanced rotation (ahead of the end of plunging) can enhance lift; however, for an insect-scale flyer, wing flexibility and passive pitching introduce aerodynamic benefits resulting in symmetric rotation and plunging to be preferred.

The advancement in micro-scale structural and material design and fabrication now enables us to engineer insect-scale flyers. Coupled with new insight gained from sensing and control research, highly controllable flight is now realizable for insect-scale flyers. For example, in a work dealing with nonwoven permeable light mats made of submicron-diameter nanofibers, Zussman et al. (2002) investigated drag in response to platform weight, average nanofiber diameter and porosity of the nonwoven mats. The results were compared to corresponding impermeable structures covered with plastic wrap. It was found that permeable platforms with holes on the order of several microns or smaller are essentially impermeable for airflow. Hence, there are many opportunities to design and optimize materials and structures to attain desirable aerodynamic performance.

Looking forward, there is a clear need to investigate sensorimotor neurobiology, musculo-skeletal mechanics and flapping, flexible wing aerodynamics, stability and control in a coherent and unified framework. In contrast to conventional fixed-wing aircraft design, with aerodynamic derivatives and control laws designed and developed separately, insect-scale flyers require that sensors, motors, muscles, skeletons, wings and bodies, and their combined behaviour and performance, be investigated altogether. Such an integrated, bio-inspired flight system will shed new light on miniaturized-scale flyers' characteristics and capabilities. There is great room for innovation in terms of developing future bio-inspired flyers capable of performing a wide range of missions, benefiting from light, flexible structures and materials and new aerodynamic understanding combined with real-time sensing and control with active and passive systems.

Besides aerial vehicles, other opportunities abound. It is estimated that 90% of the world's wild plants and 30% of the world's crops are cross-pollinated by bees, bats, birds and other insects. But as populations of bees drop due to colony collapse disorder (Lu et al. 2014), many researchers are turning to robotics as a possible alternative to help with mankind's growing food needs (Van der Schaft 2018). Swarms of autonomous robotic insects might provide a last-resort solution to a pollination crisis should we reach a critical point of mass extinction of insects (Chechetka et al. 2017; Spector 2014).

Manta rays can travel great distances, while stingrays are uniquely adapted to move on the seabed. The mesmerizing swim of both species resembles flying underwater. The knowledge gained by studying flapping flight locomotion at the insect scale can also be utilized in design of underwater fin propulsive systems mimicking rays (Chew et al. 2015; National University of Singapore 2017). About 71% of Earth's surface is covered by water. Oceans and seas have a crucially important role in our planet's ecosystem. Oceans absorb heat and stabilize temperature, and aquatic plants and animals are an essential part of the global food chain. Oceans are equally important to our economy. Surprisingly, it is estimated that humans have managed to explore only about 5% of the ocean floor. Preserving our oceans is of global strategic interest. Proper surveying and a more complete exploration of our seas and oceans is urgent for a better understanding of this complex ecosystem to help in its preservation. Autonomous bio-inspired exploration units can greatly advance the mapping of the ocean floor and help monitor marine life without disturbing the natural habitat (Frame et al. 2018; Wodinsky 2018). Other practical applications may include underwater archaeological exploration, earthquake and tsunami hazard sites evaluation and assessment of dangerous underwater sites (war wreckage).

Lightweight, self-stabilizing, flapping wing MAVs could follow atmospheric currents and be mass employed for air quality surveying and to aid meteorology studies and weather forecasting (Ristroph & Childress 2014). Besides the meteorological surveying of Earth's atmosphere, with the help of bio-inspiration and advancements of miniaturization, insect-scale MAVs are seen as a cost-effective tool for planetary exploration (Barret 2000; Hassanalian et al. 2018).

Semi-autonomous drones are used to aid the visual spectacle of dance performances and concerts, as well as aerial landscaping (Cadell 2018; Holmes 2015; Van Hemert 2015). Furthermore, international challenges and flight competitions of MAVs are a new branch of tech sports with a rapidly growing popularity that provides a continuous stimulus for development involving the public sector (Cherney 2019). Currently, these drones in practice are typically larger than the insect scale. However, the knowledge and insight gained will prove crucial for flyers focused on the current treatise due to their greater aerodynamic complexity.

References

Åke Norberg, R. (1972). The pterostigma of insect wings as an inertial regulator of wing pitch. *Journal of Comparative Physiology.* https://doi:10.1007/BF00693547

Alben, S., & Shelley, M. (2005). Coherent locomotion as an attracting state for a free flapping body. *Proceedings of the National Academy of Sciences of the United States of America*, **102**, 11163–11166.

Alben, S., Shelley, M., & Zhang, J. (2002). Drag reduction through self-similar bending of a flexible body. *Nature*, **420**(6915), 479–481.

Alexander, D. E. (1984). Unusual phase relationships between the forewings and hindwings in flying dragonflics. *Journal of Experimental Biology*, **109**, 379–383.

Alexander, D. E. (2002). *Nature's flyers: Birds, insects, and the biomechanics of flight*, 1st ed., Baltimore, MD: Johns Hopkins University Press.

Alexander, D. E. (2015). *On the wing: Insects, pterosaurs, birds, bats and the evolution of animal flight*, 1st ed., Oxford: Oxford University Press.

Alexander, R. D., & Brown Jr, W. L. (1963). Mating behavior and the origin of insect wings. *Occasional Papers of the Museum of Zoology*, **628**, 1–19.

Altshuler, D. L., Dickson, W. B., Vance, J. T., Roberts, S. P. & Dickinson, M. H. (2005). Short-amplitude high-frequency wing strokes determine the aerodynamics of honeybee flight. *Proceedings of the National Academy of Sciences of the United States of America*, **102**(50), 18213–18218.

Anderson, J. D. (2011). *Fundamentals of aerodynamics*, 5th ed., New York: McGraw Hill. https://doi:10.2514/152157

Aono, H., Liang, F., & Liu, H. (2008). Near- and far-field aerodynamics in insect hovering flight: An integrated computational study. *Journal of Experimental Biology.* https://doi:10.1242/jeb.008649

Aono, H., Shyy, W., & Liu, H. (2009). Near wake vortex dynamics of a hovering hawkmoth. *Acta Mechanica Sinica.* https://doi:10.1007/s10409-008-0210-x

Azuma, A. (2006). *The biokinetics of flying and swimming*, 2nd ed., Reston, VA: American Institute of Aeronautics and Astronautics. https://doi:10.2514/4.862502

Azuma, A., & Watanabe, T. (1988). Flight performance of a dragonfly. *Journal of Experimental Biology*, **137**, 221–252.

Babinsky, H. (2003). How do wings work? *Physics Education.* https://doi:10.1088/0031-9120/38/6/001

Barret, C. (2000). Aerobots and hydrobots for planetary exploration. In *38th Aerospace Sciences Meeting and Exhibit*, Reston, VA: American Institute of Aeronautics and Astronautics. https://doi:10.2514/6.2000-633

Bäumler, F., Gorb, S. N. & Büsse, S. (2018). Comparative morphology of the thorax musculature of adult Anisoptera (Insecta: Odonata): Functional aspects of the flight apparatus. *Arthropod Structure and Development*. https://doi:10.1016/j.asd.2018.04.003

Beem, H. R., Rival, D. E. & Triantafyllou, M. S. (2012). On the stabilization of leading-edge vortices with spanwise flow. *Experiments in Fluids*. https://doi:10.1007/s00348-011-1241-9

Bergou, A. J., Xu, S. & Wang, Z. J. (2007). Passive wing pitch reversal in insect flight. *Journal of Fluid Mechanics*. https://doi:10.1017/22112007008440

Berman, G. J., & Wang, Z. J. (2007). Energy-minimizing kinematics in hovering insect flight. *Journal of Fluid Mechanics*, **582**, 153.

Birch, J. M., & Dickinson, M. H. (2001). Spanwise flow and the attachment of the leading-edge vortex on insect wings. *Nature*. https://doi:10.1038/35089071

Birch, J. M., & Dickinson, M. H. (2003). The influence of wing-wake interactions on the production of aerodynamic forces in flapping flight. *Journal of Experimental Biology*, **206**(13), 2257–2272.

Birch, J. M., Dickson, W. B. & Dickinson, M. H. (2004). Force production and flow structure of the leading edge vortex on flapping wings at high and low Reynolds numbers. *Journal of Experimental Biology*. https://doi:10.1242/jeb.00848

Bluman, J. E., & Kang, C. (2017a). Achieving hover equilibrium in free flight with a flexible flapping wing. *Journal of Fluids and Structures*, **75**, 117–139.

Bluman, J. E., & Kang, C. (2017b). Wing-wake interaction destabilizes hover equilibrium of a flapping insect-scale wing. *Bioinspiration & Biomimetics*, **12**(4), 046004.

Bluman, J. E., Sridhar, M. K., & Kang, C. (2018). Chordwise wing flexibility may passively stabilize hovering insects. *Journal of the Royal Society Interface*, **15**(147), 20180409.

Bode-Oke, A. T., Zeyghami, S., & Dong, H. (2018). Flying in reverse: Kinematics and aerodynamics of a dragonfly in backward free flight. *Journal of the Royal Society Interface*. doi:10.1098/rsif.2018.0102

Bomphrey, R. J. (2005). The aerodynamics of Manduca sexta: Digital particle image velocimetry analysis of the leading-edge vortex. *Journal of Experimental Biology*. doi:10.1242/jeb.01471

Bomphrey, R. J., Nakata, T., Henningsson, P., & Lin, H. T. (2016). Flight of the dragonflies and damselflies. *Philosophical Transactions of the Royal Society B: Biological Sciences*. doi:10.1098/rstb.2015.0389

Brackenbury, J. H. (1994). Wing folding and free-flight kinematics in Coleoptera (Insecta): A comparative study. *Journal of Zoology*. doi:10.1111/j.1469-7998.1994.tb01572.x

Broering, T. M., Lian, Y., & Henshaw, W. (2012). Numerical investigation of energy extraction in a tandem flapping wing configuration. *AIAA Journal*. doi:10.2514/1.J051104

Bushnell, D. M., & Moore, K. J. (1991). Drag reduction in nature. *Annual Review of Fluid Mechanics*, **23**(1), 65–79.

Büsse, S., Helmker, B. & Hörnschemeyer, T. (2015). The thorax morphology of Epiophlebia (Insecta: Odonata) nymphs: Including remarks on ontogenesis and evolution. *Scientific Reports*. doi:10.1038/srep12835

Cadell, C. (2018, May). Flight of imagination: Chinese firm breaks record with 1,374 dancing drones. Reuters. Retrieved from www.reuters.com/article/us-china-drones/flight-of-imagination-chinese-firm-breaks-record-with-1374-dancing-drones-idUSKBN1I3189.

Carr, Z. R., Chen, C. & Ringuette, M. J. (2013). Finite-span rotating wings: Three-dimensional vortex formation and variations with aspect ratio. *Experiments in Fluids*, **54**(2), 1444.

Carr, Z. R., DeVoria, A. C. & Ringuette, M. J. (2015). Aspect-ratio effects on rotating wings: Circulation and forces. *Journal of Fluid Mechanics*, **767**, 497–525.

Chechetka, S. A., Yu, Y., Tange, M. & Miyako, E. (2017). Materially engineered artificial pollinators. *Chem*, **2**(2), 224–239.

Chen, Y. H., Skote, M., Zhao, Y. & Huang, W. M. (2013). Dragonfly (Sympetrum flaveolum) flight: Kinematic measurement and modelling. *Journal of Fluids and Structures*. doi:10.1016/j.jfluidstructs.2013.04.003

Chen, Y., Wang, H., Helbling, E. F. et al. (2017). A biologically inspired, flapping-wing, hybrid aerial-aquatic microrobot. *Science Robotics*, **2**(11), eaao5619.

Cheng, B., & Deng, X. (2011). Translational and rotational damping of flapping flight and its dynamics and stability at hovering. *IEEE Transactions on Robotics*, **27**(5), 849–864.

Cheng, B., Deng, X. & Hedrick, T. L. (2011). The mechanics and control of pitching manoeuvres in a freely flying hawkmoth (Manduca sexta). *Journal of Experimental Biology*, **214**(24), 4092–4106.

Cheng, B., Roll, J., Liu, Y., Troolin, D. R. & Deng, X. (2014). Three-dimensional vortex wake structure of flapping wings in hovering flight. *Journal of the Royal Society Interface*. doi:10.1098/rsif.2013.0984

Cherney, M. (2019, April). Drone racing fans have some questions: Where's the drone? Who's winning? *Wall Street Journal*. Retrieved from www.wsj.com/

articles/drone-racing-fans-have-some-questions-wheres-the-drone-whos-winning–11554305405.

Chew, C.-M., Lim, Q.-Y. & Yeo, K. S. (2015). Development of propulsion mechanism for robot manta ray. In *2015 IEEE International Conference on Robotics and Biomimetics (ROBIO)*, Washington, DC: Institute of Electrical and Electronics Engineers, pp. 1918–1923.

Chin, D. D., & Lentink, D. (2016). Flapping wing aerodynamics: From insects to vertebrates. *Journal of Experimental Biology*. doi:10.1242/jeb.042317

Chowdhury, J., Cook, L. & Ringuette, M. J. (2019). The vortex formation of an unsteady translating plate with a rotating tip. In *AIAA Scitech 2019 Forum*, Reston, VA: American Institute of Aeronautics and Astronautics. doi:10.2514/6.2019-0348

Coleman, D., & Benedict, M. (2015). Design, development and flight-testing of a robotic hummingbird. In *71st Annual Forum of the American Helicopter Society*, Virginia Beach, VA, pp. 1–18.

Combes, S. A. (2010). Materials, structure, and dynamics of insect wings as bioinspiration for MAVs. In R. Blockley & W. Shyy, eds., *Encyclopedia of Aerospace Engineering*, Hoboken, NJ: Wiley, pp. 1–10.

Combes, S. A., Crall, J. D. & Mukherjee, S. (2010). Dynamics of animal movement in an ecological context: Dragonfly wing damage reduces flight performance and predation success. *Biology Letters*, **6**(3), 426–429.

Combes, S. A., & Daniel, T. L. (2003a). Flexural stiffness in insect wings. II. Spatial distribution and dynamic wing bending. *Journal of Experimental Biology*. doi:10.1242/jeb.00524

Combes, S. A., & Daniel, T. L. (2003b). Flexural stiffness in insect wings I. Scaling and the influence of wing venation. *Journal of Experimental Biology*, **206**(17), 2979–2987.

Combes, S. A., Rundle, D. E., Iwasaki, J. M. & Crall, J. D. (2012). Linking biomechanics and ecology through predator-prey interactions: Flight performance of dragonflies and their prey. *Journal of Experimental Biology*, **215**(6), 903–913.

Cooter, R. J., & Baker, P. S. (1977). Weis-Fogh clap and fling mechanism in Locusta. *Nature*. doi:10.1038/269053a0

Darwin, C. (1859). *On the origin of the species*.

Davis, W. R. J., Kosicki, B. B., Boroson, D. M. & Kostishack, D. F. (1996). Micro air vehicles for optical surveillance. *Lincoln Laboratory Journal*.

Deng, S., Percin, M. & Van Oudheusden, B. (2015). Aerodynamic characterization of 'delfly micro' in forward flight configuration by force measurements and flow field visualization. *Procedia Engineering*, **99**, 925–929.

Deora, T., Gundiah, N. & Sane, S. P. (2017). Mechanics of the thorax in flies. *Journal of Experimental Biology.* doi:10.1242/jeb.128363

Desbiens, A. L., Chen, Y. & Wood, R. J. (2013). A wing characterization method for flapping-wing robotic insects. *IEEE International Conference on Intelligent Robots and Systems*, Washington, DC: Institute of Electrical and Electronics Engineers, pp. 1367–1373.

Dewey, P. A., Boschitsch, B. M., Moored, K. W., Stone, H. A. & Smits, A. J. (2013). Scaling laws for the thrust production of flexible pitching panels. *Journal of Fluid Mechanics*, **732**, 29–46.

Dickinson, M. H. (1999). Haltere-mediated equilibrium reflexes of the fruit fly, Drosophila melanogaster. *Philosophical Transactions of the Royal Society B: Biological Sciences.* doi:10.1098/rstb.1999.0442

Dickinson, M. H., & Gotz, K. G. (1993). Unsteady aerodynamic performance of model wings at low Reynolds numbers. *Journal of Experimental Biology*, **174**(1), 45–64.

Dickinson, M. H., Lehmann, F. O. & Gotz, K. G. (1993). The active control of wing rotation by Drosophila. *Journal of Experimental Biology.*

Dickinson, M. H., Lehmann, F.-O. & Sane, S. P. (1999). Wing rotation and the aerodynamic basis of insect flight. *Science.* doi:10.1126/science.284.5422.1954

Dudley, R. (2000). *The biomechanics of insect flight: Form, function, evolution*, Princeton, NJ: Princeton University Press.

Eldredge, J. D., Toomey, J. & Medina, A. (2010). On the roles of chord-wise flexibility in a flapping wing with hovering kinematics. *Journal of Fluid Mechanics*, **659**, 94–115.

Ellington, C. P. (1984a). The aerodynamics of hovering insect flight. I. The quasi-steady analysis. *Philosophical Transactions of the Royal Society B: Biological Sciences*, **305**(1122), 1–15.

Ellington, C. P. (1984b). The aerodynamics of hovering insect flight. II. Morphological parameters. *Philosophical Transactions of the Royal Society B: Biological Sciences*, **305**(1122), 17–40.

Ellington, C. P. (1984c). The aerodynamics of hovering insect flight. III. Kinematics. *Philosophical Transactions of the Royal Society B: Biological Sciences*, **305**(1122), 41–78.

Ellington, C. P., Van Berg, C., Den Willmott, A. P., & Thomas, A. L. R. (1996). Leading-edge vortices in insect flight. *Nature.* doi:10.1038/384626a0

Ennos, A. R. R. (1988). The inertial cause of wing rotation in Diptera. *Journal of Experimental Biology*, **140**(1), 161–169.

Ennos, A. R. R. (1989). Inertial and aerodynamic torques on the wings of diptera in flight. *Journal of Experimental Biology*, **142**, 87–95.

Faruque, I., & Humbert, J. S. (2010). Dipteran insect flight dynamics. Part 1: longitudinal motion about hover. *Journal of Theoretical Biology*, **264**(2), 538–552.

Fédrigo, O., & Wray, G. A. (2010). Developmental evolution: How beetles evolved their shields. *Current Biology*. doi:10.1016/j.cub.2009.12.012

Forbes, W. T. M. (1943). The origin of wings and venational types in insects. *American Midland Naturalist*, **29**(2), 381.

Frame, J., Lopez, N., Curet, O. & Engeberg, E. D. (2018). Thrust force characterization of free-swimming soft robotic jellyfish. *Bioinspiration & Biomimetics*, **13**(6), 064001.

Fry, S. N., Sayaman, R. & Dickinson, M. H. (2005). The aerodynamics of hovering flight in Drosophila. *Journal of Experimental Biology*, **208**(12), 2303–2318.

Fu, J., Hefler, C., Qiu, H. H. & Shyy, W. (2014). Effects of aspect ratio on flapping wing aerodynamics in animal flight. *Acta Mechanica Sinica*, **30**(6), 776–786.

Fu, J., Liu, X., Shyy, W. & Qiu, H. H. (2018). Effects of flexibility and aspect ratio on the aerodynamic performance of flapping wings. *Bioinspiration & Biomimetics*. doi:10.1088/1748-3190/aaaac1

Fu, J., Shyy, W. & Qiu, H. H. (2017). Effects of aspect ratio on vortex dynamics of a rotating wing. *AIAA Journal*. doi:10.2514/1.j055764

Garmann, D. J., & Visbal, M. R. (2014). Dynamics of revolving wings for various aspect ratios. *Journal of Fluid Mechanics*, **748**, 932–956.

Gegenbaur, C., Bell, F. J. & Lankester, E. R. (1878). *Elements of comparative anatomy*, 2nd ed., London: Macmillan.

Ghiradella, H. (1994). Structure of butterfly scales: Patterning in an insect cuticle. *Microscopy Research and Technique*. doi:10.1002/jemt.1070270509

Ghiradella, H. (1998). Hairs, bristles, and scales. In *Microscopic anatomy of invertebrates, Volume 11A, Insecta*.

Gordnier, R. E., & Attar, P. J. (2014). Impact of flexibility on the aerodynamics of an aspect ratio two membrane wing. *Journal of Fluids and Structures*, **45** (February), 138–152.

Gordnier, R. E., Kumar Chimakurthi, S., Cesnik, C. E. S. et al. (2013). High-fidelity aeroelastic computations of a flapping wing with spanwise flexibility. *Journal of Fluids and Structures*, **40**(July), 86–104.

Guizzo, E. (2011, April). Robotic aerial vehicle captures dramatic footage of Fukushima reactors. IEEE Spectrum. Retrieved from https://spectrum.ieee .org/automaton/robotics/industrial-robots/robotic-aerial-vehicle-at-fukush ima-reactors

Gunnell, G. F., & Simmons, N. B., eds. (2012). *Evolutionary history of bats*, Cambridge: Cambridge University Press. doi:10.1017/CBO97811390 45599

Haas, F., & Wootton, R. J. (1996). Two basic mechanisms in insect wing folding. *Proceedings of the Royal Society B: Biological Sciences*. doi:10.1098/rspb.1996.0241

Habib, M. B. (2013). Capacity for the cretaceous pterosaur Anhanguera to launch from water. *FASEB Journal*.

Han, J.-S., Chang, J. W. & Cho, H.-K. (2015). Vortices behavior depending on the aspect ratio of an insect-like flapping wing in hover. *Experiments in Fluids*, **56**(9), 181.

Harbig, R. R., Sheridan, J. & Thompson, M. C. (2014). The role of advance ratio and aspect ratio in determining leading-edge vortex stability for flapping flight. *Journal of Fluid Mechanics*, **751**, 71–105.

Hassanalian, M., Rice, D. & Abdelkefi, A. (2018). Evolution of space drones for planetary exploration: A review. *Progress in Aerospace Sciences*, **97**, 61–105.

Heathcote, S., & Gursul, I. (2007). Flexible flapping airfoil propulsion at low Reynolds numbers. *AIAA Journal*, **45**(5), 1066–1079.

Heathcote, S., Martin, D. & Gursul, I. (2004). Flexible flapping airfoil propulsion at zero freestream velocity. *AIAA Journal*. doi:10.2514/1.5299

Heathcote, S., Wang, Z. & Gursul, I. (2008). Effect of spanwise flexibility on flapping wing propulsion. *Journal of Fluids and Structures*. doi:10.1016/j .jfluidstructs.2007.08.003

Hedenström, A., & Johansson, L. C. (2015). Bat flight. *Current Biology*. doi:10.1016/j.cub.2015.04.002

Hedrick, T. L. (2011). Damping in flapping flight and its implications for manoeuvring, scaling and evolution. *Journal of Experimental Biology*, **214**(24), 4073–4081.

Hedrick, T. L., Cheng, B. & Deng, X. (2009). Wingbeat time and the scaling of passive rotational damping in flapping flight. *Science*, **324** (5924), 252.

Hedrick, T. L., Combes, S. A. & Miller, L. A. (2015). Recent developments in the study of insect flight. *Canadian Journal of Zoology*, **93**(12), 925–943.

Hefler, C., Noda, R., Qiu, H. H. & Shyy, W. (2020). Aerodynamic performance of a free-flying dragonfly: A span-resolved investigation. *Physics of Fluids*. doi:10.1063/1.5145199

Hefler, C., Qiu, H. H. & Shyy, W. (2018). Aerodynamic characteristics along the wing span of a dragonfly Pantala flavescens. *Journal of Experimental Biology*, **221**(19). doi:10.1242/jeb.171199

Holmes, K. (2015, December). A self-organizing drone army dances with humans. VICE. Retrieved from www.vice.com/en_us/article/pgqxj9/col lmot-robotics-drone-dance

Hone, D. W. E., Van Rooijen, M. K. & Habib, M. B. (2015). The wingtips of the pterosaurs: anatomy, aeronautical function and ecological implications. *Palaeogeography, Palaeoclimatology, Palaeoecology.* doi:10.1016/j .palaeo.2015.08.046

Hsieh, C. T., Kung, C. F., Chang, C. C. & Chu, C. C. (2010). Unsteady aerodynamics of dragonfly using a simple wing-wing model from the perspective of a force decomposition. *Journal of Fluid Mechanics.* doi:10.1017/ S0022112010003484

Hu, Z., & Deng, X. Y. (2014). Aerodynamic interaction between forewing and hindwing of a hovering dragonfly. *Acta Mechanica Sinica.* doi:10.1007/ s10409-014-0118-6

Huang, H., & Sun, M. (2007). Dragonfly forewing-hindwing interaction at various flight speeds and wing phasing. *AIAA Journal.* doi:10.2514/ 1.24666

Ishihara, D., Horie, T. & Denda, M. (2009a). A two-dimensional computational study on the fluid-structure interaction cause of wing pitch changes in dipteran flapping flight. *Journal of Experimental Biology.* doi:10.1242/ jeb.020404

Ishihara, D., Yamashita, Y., Horie, T., Yoshida, S. & Niho, T. (2009b). Passive maintenance of high angle of attack and its lift generation during flapping translation in crane fly wing. *Journal of Experimental Biology,* **212**(23), 3882–3891.

Jafferis, N. T., Helbling, E. F., Karpelson, M. & Wood, R. J. (2019). Untethered flight of an insect-sized flapping-wing microscale aerial vehicle. *Nature,* **570**(7762), 491–495.

Jones, K. D., Lund, T. C. & Platzer, M. F. (2002). Experimental and computational investigation of flapping wing propulsion for micro air vehicles. *Progress in Astronautics and Aeronautics,* **195**, 307–340.

Kang, C., Aono, H., Cesnik, C. E. S. & Shyy, W. (2011). Effects of flexibility on the aerodynamic performance of flapping wings. *Journal of Fluid Mechanics,* **689**, 32–74.

Kang, C., Cranford, J., Sridhar, M. K., Kodali, D., Landrum, D. B. & Slegers, N. (2018). Experimental characterization of a butterfly in climbing flight. *AIAA Journal,* **56**(1), 15–24.

Kang, C., & Shyy, W. (2013). Scaling law and enhancement of lift generation of an insect-size hovering flexible wing. *Journal of the Royal Society Interface,* **10**(85), 20130361.

Kang, C., & Shyy, W. (2014). Analytical model for instantaneous lift and shape deformation of an insect-scale flapping wing in hover. *Journal of the Royal Society Interface*, **11**(101), 20140933.

Karásek, M., Muijres, F. T., De Wagter, C., Remes, B. D. W. & de Croon, G. C. H. E. (2018). A tailless aerial robotic flapper reveals that flies use torque coupling in rapid banked turns. *Science*, **361**(6407), 1089–1094.

Katz, J., & Weihs, D. (1978). Hydrodynamic propulsion by large amplitude oscillation of an airfoil with chordwise flexibility. *Journal of Fluid Mechanics*. doi:10.1017/S0022112078002220

Keennon, M., Klingebiel, K., Won, H. & Andriukov, A. (2012). Development of the nano hummingbird: A tailless flapping wing micro air vehicle. In *50th AIAA Aerospace Sciences Meeting AIAA 2012–0588*, American Institute of Aeronautics and Astronautics. doi:10.2514/6.2012-588

Kodali, D., Medina, C., Kang, C. & Aono, H. (2017). Effects of spanwise flexibility on the performance of flapping flyers in forward flight. *Journal of the Royal Society Interface*, **14**(136), 20170725.

Kruyt, J. W., Quicazán-Rubio, E. M., Van Heijst, G. F., Altshuler, D. L. & Lentink, D. (2014). Hummingbird wing efficacy depends on aspect ratio and compares with helicopter rotors. *Journal of the Royal Society Interface*, **11**(99), 20140585.

Kruyt, J. W., van Heijst, G. F., Altshuler, D. L. & Lentink, D. (2015). Power reduction and the radial limit of stall delay in revolving wings of different aspect ratio. *Journal of the Royal Society Interface*, **12**(105), 20150051.

Kumar, A., Kumar, N., Das, R., Lakhani, P. & Bhushan, B. (2019). In vivo structural dynamic analysis of the dragonfly wing: The effect of stigma as its modulator. *Philosophical Transactions of the Royal Society A: Mathematical, Physical and Engineering Sciences*. doi:10.1098/rsta.2019.0132

Lai, G., & Shen, G. (2012). Experimental investigation on the wing-wake interaction at the mid stroke in hovering flight of dragonfly. *Science China: Physics, Mechanics and Astronomy*. doi:10.1007/s11433-012-4907-2

Lan, S., & Sun, M. (2001). Aerodynamic force and flow structures of two airfoils in flapping motions. *Acta Mechanica Sinica*, **17**(4), 310–331.

Lee, Y. J., Lua, K. B. & Lim, T. T. (2016). Aspect ratio effects on revolving wings with Rossby number consideration. *Bioinspiration & Biomimetics*, **11**(5). doi:10.1088/1748-3190/11/5/056013

Lehmann, F. O. (2004). The mechanisms of lift enhancement in insect flight. *Naturwissenschaften*. doi:10.1007/s00114-004-0502-3

Lehmann, F.-O. (2008). When wings touch wakes: Understanding locomotor force control by wake-wing interference in insect wings. *Journal of Experimental Biology*. doi:10.1242/jeb.007575

Lehmann, F.-O., & Dickinson, M. H. (1997). The changes in power requirements and muscle efficiency during elevated force production in the fruit fly Drosophila melanogaster. *Journal of Experimental Biology*, **200**(7), 1133–1143.

Lehmann, F.-O., Sane, S. P. & Michael, D. (2005). The aerodynamic effects of wing-wing interaction in flapping insect wings. *Journal of Experimental Biology*, **208**(16), 3075–3092.

Lentink, D., & Dickinson, M. H. (2009). Rotational accelerations stabilize leading edge vortices on revolving fly wings. *Journal of Experimental Biology*. doi:10.1242/jeb.022269

Lentink, D., Jongerius, S. R. & Bradshaw, N. L. (2010). The scalable design of flapping micro-air vehicles inspired by insect flight. *Flying Insects and Robots*. doi:10.1007/978-3-540-89393-6_14

Lian, Y., Broering, T., Hord, K. & Prater, R. (2014). The characterization of tandem and corrugated wings. *Progress in Aerospace Sciences*. doi:10.1016/j.paerosci.2013.08.001

Liang, B., & Sun, M. (2013). Nonlinear flight dynamics and stability of hovering model insects. *Journal of the Royal Society Interface*, **10**(85), 20130269.

Liu, H., Ravi, S., Kolomenskiy, D. & Tanaka, H. (2016). Biomechanics and biomimetics in insect-inspired flight systems. *Philosophical Transactions of the Royal Society B: Biological Sciences*, **371**(1704), 20150390.

Lu, C., Warchol, K. M. & Callahan, R. A. (2014). Sub-lethal exposure to neonicotinoids impaired honey bees winterization before proceeding to colony collapse disorder. *Bulletin of Insectology*, **67**(1), 125–130.

Lucas, K. N., Thornycroft, P. J. M., Gemmell, B. J., Colin, S. P., Costello, J. H. & Lauder, G. V. (2015). Effects of non-uniform stiffness on the swimming performance of a passively-flexing, fish-like foil model. *Bioinspiration & Biomimetics*, **10**(5), 056019.

Luo, G., & Sun, M. (2005). Effects of corrugation and wing planform on the aerodynamic force production of sweeping model insect wings. *Acta Mechanica Sinica*, **21**, 531–541.

Ma, K. Y., Chirarattananon, P., Fuller, S. B. & Wood, R. J. (2013). Controlled flight of a biologically inspired, insect-scale robot. *Science*, **340**(6132), 603–607.

MacRae, M. (2016). 5 new applications for drones . ASME. Retrieved August 3, 2019, from www.asme.org/topics-resources/content/5-new-applications-for-drones

Mahardika, N., Viet, N. Q. & Park, H. C. (2011). Effect of outer wing separation on lift and thrust generation in a flapping wing system. *Bioinspiration & Biomimetics*. doi:10.1088/1748-3182/6/3/036006

Marden, J. H. (1987). Maximum lift production during takeoff in flying animals. *Journal of Experimental Biology.*

Maybury, W. J., & Lehmann, F.-O. (2004). The fluid dynamics of flight control by kinematic phase lag variation between two robotic insect wings. *Journal of Experimental Biology.* doi:10.1242/jeb.01319

Mcmichael, J. M., & Francis, M. S. (1997). Micro air vehicles: Toward a new dimension in flight. *Unmanned Systems.*

Michelin, S., & Smith, S. G. L. (2009). Resonance and propulsion performance of a heaving flexible wing. *Physics of Fluids*, **21**, 71902.

Miller, L. A., & Peskin, C. S. (2005). A computational fluid dynamics of 'clap and fling' in the smallest insects. *Journal of Experimental Biology.* doi:10.1242/jeb.01376

Misof, B., Liu, S., Meusemann, K. et al. (2014). Phylogenomics resolves the timing and pattern of insect evolution. *Science*, **346**(6210), 763–767.

Mountcastle, A. M., & Combes, S. A. (2013). Wing flexibility enhances load-lifting capacity in bumblebees. *Proceedings of the Royal Society B: Biological Sciences*, **280**(1759), 20130531.

Mountcastle, A. M., & Combes, S. A. (2014). Biomechanical strategies for mitigating collision damage in insect wings: Structural design versus embedded elastic materials. *Journal of Experimental Biology*, **217**(7), 1108–1115.

Muijres, F. T., Johansson, L. C., Barfield, R., Wolf, M., Spedding, G. R. & Hedenström, A. (2008). Leading-edge vortex improves lift in slow-flying bats. *Science.* doi:10.1126/science.1153019

Müller, F. (1877). Ueber haarpinsel, filzflecke und ähnliche gebilde auf den flügeln männlicher schmetterlinge. *Jenaische Zeitschrift Für Naturwissenschaft*, **11**, 99–114.

Muniappan, A., Baskar, V. & Duriyanandhan, V. (2005). Lift and thrust characteristics of flapping wing micro air vehicle (MAV). In *43rd AIAA Aerospace Sciences Meeting and Exhibit*, Reston, VA: American Institute of Aeronautics and Astronautics. doi:10.2514/6.2005-1055

Munson, B. R., Rothmayer, A. P., Okiishi, T. H. & Huebsch, W. W. (2012). *Fundamentals of fluid mechanics*, 7th ed., Hoboken, NJ: Wiley.

Nakata, T., & Liu, H. (2012). Aerodynamic performance of a hovering hawk-moth with flexible wings: a computational approach. *Proceedings. Biological Sciences / The Royal Society*, **279**(1729), 722–731.

National University of Singapore. (2017, November). NUS-developed manta ray robot swims faster and operates up to 10 hours. *NUS News: NUS Media Relations Team.* Retrieved from https://phys.org/news/2017-11-nus-developed-manta-ray-robot-faster.html.

Neville, A. C. (1960). Aspects of flight mechanics in anisopterous dragonflies. *Journal of Experimental Biology.*

Norberg, R. Å. (1975). Hovering flight of the dragonfly Aeschna juncea L., kinematics and aerodynamics. *Swimming and Flying in Nature.* doi:10.1007/978-1-4757-1326-8_19

Norberg, U. M. (1990). *Vertebrate flight: Mechanics, physiology, morphology, ecology and evolution,* Berlin: Springer.

Norberg, U. M. (2002). Evolution of vertebrate flight: An aerodynamic model for the transition from gliding to active flight. *American Naturalist.* doi:10.1086/284419

Novacek, M. J. (1985). Evidence for echolocation in the oldest known bats. *Nature.* doi:10.1038/315140a0

Orlowski, C. T., & Girard, A. R. (2012a). Dynamics, stability, and control analyses of flapping wing micro-air vehicles. *Progress in Aerospace Sciences,* **51**, 18–30.

Orlowski, C. T., & Girard, A. R. (2012b). Longitudinal flight dynamics of flapping-wing micro air vehicles. *Journal of Guidance, Control, and Dynamics,* **35**(4), 1115–1131.

Ozen, C., & Rockwell, D. (2013). Flow structure on a rotating wing: Effect of wing aspect ratio and shape. In *51st AIAA Aerospace Sciences Meeting including the New Horizons Forum and Aerospace Exposition, Reston, Virginia, 7 – 10 January 2013,* Grapevine, TX: American Institute of Aeronautics and Astronautics. doi:10.2514/6.2013-676

Park, H., & Choi, H. (2012). Kinematic control of aerodynamic forces on an inclined flapping wing with asymmetric strokes. *Bioinspiration and Biomimetics.* doi:10.1088/1748-3182/7/1/016008

Park, J. H., & Yoon, K. J. (2008). Designing a biomimetic ornithopter capable of sustained and controlled flight. *Journal of Bionic Engineering.* doi:10.1016/S1672-6529(08)60005-0

Parle, E., Dirks, J. H. & Taylor, D. (2017). Damage, repair and regeneration in insect cuticle: The story so far, and possibilities for the future. *Arthropod Structure and Development.* doi:10.1016/j.asd.2016.11.008

Pass, G. (2018). Beyond aerodynamics: The critical roles of the circulatory and tracheal systems in maintaining insect wing functionality. *Arthropod Structure and Development.* doi:10.1016/j.asd.2018.05.004

Pennycuick, C. J. (2008). *Modelling the flying bird.* Theoretical Ecology Series. Burlington, MA: Academic Press. doi:10.1007/s13398-014-0173-7.2

Percin, M., Hu, Y., Van Oudheusden, B. W., Remes, B. & Scarano, F. (2011). Wing flexibility effects in clap-and-fling. *International Journal of Micro Air Vehicles,* **3**(4), 217–227.

Phillips, N., Knowles, K. & Bomphrey, R. J. (2015). The effect of aspect ratio on the leading-edge vortex over an insect-like flapping wing. *Bioinspiration & Biomimetics*, **10**(5), 056020.

Platzer, M. F., Jones, K. D., Young, J. & Lai, J. C. S. (2008). Flapping-wing aerodynamics: Progress and challenges. *AIAA Journal*. doi:10.2514/1.29263

Pornsin-Sirirak, T. N., Tai, Y.-C., Ho, C.-M. & Keennon, M. (2001). Microbat: A palm-sized electrically powered ornithopter. *Proceedings of NASA/JPL Workshop on Biomorphic Robotics*.

Rajabi, H., Rezasefat, M., Darvizeh, A. et al. (2016). A comparative study of the effects of constructional elements on the mechanical behaviour of dragonfly wings. *Applied Physics A: Materials Science and Processing*. doi:10.1007/s00339-015-9557-6

Ramamurti, R., & Sandberg, W. (2001). Simulation of flow about flapping airfoils using finite element incompressible flow solver. *AIAA Journal*, **39** (2), 253–260.

Ramamurti, R., & Sandberg, W. (2007). A computational investigation of the three-dimensional unsteady aerodynamics of Drosophila hovering and maneuvering. *Journal of Experimental Biology*. doi:10.1242/jeb.02704

Ramananarivo, S., Godoy-Diana, R. & Thiria, B. (2011). Rather than resonance, flapping wing flyers may play on aerodynamics to improve performance. *Proceedings of the National Academy of Sciences of the United States of America*, **108**(15), 5964–5969.

Reavis, M., & Luttges, M. (1988). Aerodynamic forces produced by a dragonfly. In *26th Aerospace Sciences Meeting*, Reston, VA: American Institute of Aeronautics and Astronautics. doi:10.2514/6.1988-330

Richardson, P. L. (2011). How do albatrosses fly around the world without flapping their wings? *Progress in Oceanography*. doi:10.1016/j.pocean.2010.08.001

Riggs, P., Bowyer, A. & Vincent, J. (2010). Advantages of a biomimetic stiffness profile in pitching flexible fin propulsion. *Journal of Bionic Engineering*. doi:10.1016/S1672-6529(09)60203-1

Ristroph, L., & Childress, S. (2014). Stable hovering of a jellyfish-like flying machine. *Journal of the Royal Society Interface*. doi:10.1098/rsif.2013.0992

Rival, D., Schönweitz, D. & Tropea, C. (2011). Vortex interaction of tandem pitching and plunging plates: A two-dimensional model of hovering dragonfly-like flight. *Bioinspiration & Biomimetics*, **6**(1), 016008.

Roccia, B. A., Preidikman, S., Verstraete, M. L. & Mook, D. T. (2017). Influence of spanwise twisting and bending on lift generation in MAV-like flapping wings. *Journal of Aerospace Engineering*, **30**(1), 04016079.

Rudolph, R. (1976a). Preflight behaviour and the initiation of flight in tethered and unrestrained dragonfly, Calopteryx splendens (Harris) (Zygoptera: Calopterygidae). *Odonatologica*, **5**(1), 59–64.

Rudolph, R. (1976b). Some aspects of wing kinematics in Calopteryx splendens (Harris) (Zygoptera: Calopterygidae). *Odonatologica*, **5**(2), 119–127.

Rüppel, G. (1989). Kinematic analysis of symmetrical flight manoeuvres of Odonata. *Journal of Experimental Biology*, **144**(1).

Ryu, Y., Chang, J. W. & Chung, J. (2016). Aerodynamic force and vortex structures of flapping flexible hawkmoth-like wings. *Aerospace Science and Technology*, **56**, 183–196.

Salami, E., Ward, T. A., Montazer, E. & Ghazali, N. N. N. (2019). A review of aerodynamic studies on dragonfly flight. *Proceedings of the Institution of Mechanical Engineers, Part C: Journal of Mechanical Engineering Science*. doi:10.1177/0954406219861133

Salumäe, T., & Kruusmaa, M. (2011). A flexible fin with bio-inspired stiffness profile and geometry. *Journal of Bionic Engineering*. doi:10.1016/S1672-6529(11)60047-4

Sane, S. P. (2003). The aerodynamics of insect flight. *Journal of Experimental Biology*, **206**(23), 4191–4208.

Sane, S. P., & Dickinson, M. H. (2001). The control of flight force by a flapping wing: Lift and drag production. *Journal of Experimental Biology*.

Sane, S. P., & Dickinson, M. H. (2002). The aerodynamic effects of wing rotation and a revised quasi-steady model of flapping flight. *Journal of Experimental Biology*.

Santhanakrishnan, A., Robinson, A. K., Jones, S. et al. (2014). Clap and fling mechanism with interacting porous wings in tiny insect flight. *Journal of Experimental Biology*. doi:10.1242/jeb.084897

Schmidt, J., O'Neill, M., Dirks, J. H. & Taylor, D. (2020). An investigation of crack propagation in an insect wing using the theory of critical distances. *Engineering Fracture Mechanics*. doi:10.1016/j.engfracmech.2020.107052

Shang, J. K., Combes, S. A., Finio, B. M. & Wood, R. J. (2009). Artificial insect wings of diverse morphology for flapping-wing micro air vehicles. *Bioinspiration & Biomimetics*, **4**(3), 36002.

Shelley, M. J., & Zhang, J. (2011). Flapping and bending bodies interacting with fluid flows. *Annual Review of Fluid Mechanics*, **43**(1), 449–465.

Shyy, W., Aono, H., Chimakurthi, S. K. et al. (2010). Recent progress in flapping wing aerodynamics and aeroelasticity. *Progress in Aerospace Sciences*, **46**(7), 284–327.

Shyy, W., Aono, H., Kang, C., & Liu, H. (2013). *An introduction to flapping wing aerodynamics*, Cambridge: Cambridge University Press. doi:10.1017/CBO9781139583916

Shyy, W., Berg, M. & Ljungqvist, D. (1999). Flapping and flexible wings for biological and micro air vehicles. *Progress in Aerospace Sciences*, **35**(5), 455–505.

Shyy, W., Ifju, P. & Viieru, D. (2005). Membrane wing-based micro air vehicles. *Applied Mechanics Reviews*, **58**(4), 283–301.

Shyy, W., Kang, C. K., Chirarattananon, P., Ravi, S. & Liu, H. (2016). Aerodynamics, sensing and control of insect-scale flapping-wing flight. *Proceedings of the Royal Society A: Mathematical, Physical and Engineering Sciences*. doi:10.1098/rspa.2015.0712

Shyy, W., Lian, Y., Tang, J. et al. (2008). Computational aerodynamics of low Reynolds number plunging, pitching and flexible wings for MAV applications. *Acta Mechanica Sinica/Lixue Xuebao*. doi:10.1007/s10409-008-0164-z

Shyy, W., Lian, Y., Tang, J., Viieru, D., & Liu, H. (2007). *Aerodynamics of low Reynolds number flyers*, Cambridge: Cambridge University Press. doi:10.1017/CBO9780511551154

Shyy, W., & Liu, H. (2007). Flapping wings and aerodynamic lift: The role of leading-edge vortices. *AIAA Journal*, **45**(12), 2817–2819.

Shyy, W., Trizila, P., Kang, C.-K. & Aono, H. (2009). Can tip vortices enhance lift of a flapping wing? *AIAA Journal*. doi:10.2514/1.41732

Somps, C., & Luttges, M. (1985). Dragonfly flight: Novel uses of unsteady separated flows. *Science*. doi:10.1126/science.228.4705.1326

Somps, C., & Luttges, M. (1986). Response: Dragonfly aerodynamics. *Science*, **231**(4733), 10.

Spagnolie, S., Moret, L., Shelley, M. J. & Zhang, J. (2010). Surprising behaviors in flapping locomotion with passive pitching. *Physics of Fluids*, **22**, 41903.

Spector, D. (2014, July). Harvard robobees closer to pollinating crops instead of real bees. *Business Insider*. Retrieved from www.businessinsider.com/harvard-robobees-closer-to-pollinating-crops-2014–6

Sridhar, M. K., & Kang, C. (2015). Aerodynamic performance of two-dimensional, chordwise flexible flapping wings at fruit fly scale in hover flight. *Bioinspiration & Biomimetics*, **10**(3), 036007.

Srygley, R. B., & Thomas, A. L. R. (2002). Unconventional lift-generating mechanisms in free-flying butterflies. *Nature*. doi:10.1038/nature01223

Su, W., & Cesnik, C. E. S. (2011). Flight dynamic stability of a flapping wing micro air vehicle in hover. In *Collection of Technical Papers: AIAA/ASME/*

ASCE/AHS/ASC Structures, Structural Dynamics and Materials Conference, Denver, CO.

Sullivan, T. N., Wang, B., Espinosa, H. D., & Meyers, M. A. (2017). Extreme lightweight structures: Avian feathers and bones. *Materials Today*. doi:10.1016/j.mattod.2017.02.004

Sun, M. (2014). Insect flight dynamics: Stability and control. *Reviews of Modern Physics*, **86**(2), 615–646.

Sun, M., & Lan, S. L. (2004). A computational study of the aerodynamic forces and power requirements of dragonfly (Aeschna juncea) hovering. *Journal of Experimental Biology*. doi:10.1242/jeb.00969

Sun, M., & Tang, J. (2002). Unsteady aerodynamic force generation by a model fruit fly wing in flapping motion. *Journal of Experimental Biology*.

Sun, M., Wang, J. K. & Xiong, Y. (2007). Dynamic flight stability of hovering insects. *Acta Mechanica Sinica*, **23**(3), 231–246.

Sun, M., & Xiong, Y. (2005). Dynamic flight stability of a hovering bumblebee. *Journal of Experimental Biology*, **208**(3), 447–459.

Sunada, S., Zeng, L. & Kawachi, K. (1998). The relationship between dragonfly wing structure and torsional deformation. *Journal of Theoretical Biology*, **193**(1), 39–45.

Sutherland, B., ed. (2011). *The economist: Modern warfare, intelligence and deterrence*, London: Economist Group Press. Retrieved from https://profile books.com/the-economist-modern-warfare-intelligence-and-deterrence-ebook.html.

Taha, H. E., Hajj, M. R. & Nayfeh, A. H. (2012). Flight dynamics and control of flapping-wing MAVs: A review. *Nonlinear Dynamics*, **70**(2), 907–939.

Taha, H. E., Hajj, M. R. & Nayfeh, A. H. (2014). Longitudinal flight dynamics of hovering MAVs/insects. *Journal of Guidance, Control, and Dynamics*, **37**(3), 970–979.

Taha, H. E., Tahmasian, S., Woolsey, C. A., Nayfeh, A. H. & Hajj, M. R. (2015). The need for higher-order averaging in the stability analysis of hovering, flapping-wing flight. *Bioinspiration & Biomimetics*, **10**(1), 016002.

Taira, K., & Colonius, T. (2009). Three-dimensional flows around low-aspect-ratio flat-plate wings at low Reynolds numbers. *Journal of Fluid Mechanics*. doi:10.1017/S0022112008005314

Tay, W. B., Van Oudheusden, B. W. & Bijl, H. (2015). Numerical simulation of a flapping four-wing micro-aerial vehicle. *Journal of Fluids and Structures*, **55**, 237–261.

Taylor, G. K., Nudds, R. L. & Thomas, A. L. R. (2003). Flying and swimming animals cruise at a Strouhal number tuned for high power efficiency. *Nature*. doi:10.1038/nature02000

Taylor, G. K., & Thomas, A. L. R. (2002). Animal flight dynamics. II. Longitudinal stability in flapping flight. *Journal of Theoretical Biology*, **214**, 351–370.

Thiria, B., & Godoy-Diana, R. (2010). How wing compliance drives the efficiency of self-propelled flapping flyers. *Physical Review E*, **82**(1), 15303.

Thomas, A. L. R., Taylor, G. K., Srygley, R. B., Nudds, R. L., & Bomphrey, R. J. (2004). Dragonfly flight: Free-flight and tethered flow visualizations reveal a diverse array of unsteady lift-generating mechanisms, controlled primarily via angle of attack. *Journal of Experimental Biology*. doi:10.1242/jeb.01262

Triantafyllou, M. S., Triantafyllou, G. S. & Yue, D. K. P. (2000). Hydrodynamics of fishlike swimming. *Annual Review of Fluid Mechanics*. doi:10.1146/annurev.fluid.32.1.33

Trizila, P., Kang, C., Aono, H., Shyy, W. & Visbal, M. (2011). Low-Reynolds-number aerodynamics of a flapping rigid flat plate. *AIAA Journal*, **49**(4), 806–823.

Tu, Z., Fei, F. & Deng, X. (2020a). Untethered flight of an at-scale dual-motor hummingbird robot with bio-inspired decoupled wings. *IEEE Robotics and Automation Letters*, **5**(3), 4194–4201.

Tu, Z., Fei, F., Zhang, J. & Deng, X. (2020b). An at-scale tailless flapping-wing hummingbird robot. I. Design, optimization, and experimental validation. *IEEE Transactions on Robotics*, **36**(5), 1511–1525.

Usherwood, J. R., & Ellington, C. P. (2002). The aerodynamics of revolving wings II. Propeller force coefficients from mayfly to quail. *Journal of Experimental Biology*, **205**(11), 1565–76.

Usherwood, J. R., & Lehmann, F.-O. (2008). Phasing of dragonfly wings can improve aerodynamic efficiency by removing swirl. *Journal of the Royal Society Interface*. doi:10.1098/rsif.2008.0124

Van den Berg, C., & Ellington, C. P. (1997). The three-dimensional leading-edge vortex of a 'hovering' model hawkmoth. *Philosophical Transactions of the Royal Society B: Biological Sciences*. doi:10.1098/rstb.1997.0024

Van der Schaft, P. (2018, March). Pollination drones seen as assistants for ailing bees. *Robotics Business Review*. Retrieved from www.roboticsbusinessreview.com/agriculture/pollination-drones-assist-ailing-bees/.

Van Hemert, K. (2015, March). The shadow ballet where drones dance with humans. *WIRED*. Retrieved from www.wired.com/2015/03/shadow-ballet-drones-dance-humans/.

Vanella, M., Fitzgerald, T., Preidikman, S., Balaras, E. & Balachandran, B. (2009). Influence of flexibility on the aerodynamic performance of a hovering wing. *Journal of Experimental Biology*, **212**(1), 95–105.

Vargas, A., Mittal, R. & Dong, H. (2008). A computational study of the aerodynamic performance of a dragonfly wing section in gliding flight. *Bioinspiration & Biomimetics*, **3**(2), 026004.

Viscor, G., & Fuster, J. F. (1987). Relationships between morphological parameters in birds with different flying habits. *Comparative Biochemistry and Physiology – Part A: Physiology*. doi:10.1016/0300-9629(87)90118-6

Vukusic, P., & Sambles, J. R. (2001). Shedding light on butterfly wings. *Physics, Theory, and Applications of Periodic Structures in Optics*. doi:10.1117/12.451481

Wakeling, J., & Ellington, C. (1997). Dragonfly flight. II. Velocities, accelerations and kinematics of flapping flight. *Journal of Experimental Biology*, **200**(3), 557–582.

Walker, S. M., Thomas, A. L. R. & Taylor, G. K. (2009). Deformable wing kinematics in the desert locust: How and why do camber, twist and topography vary through the stroke? *Journal of the Royal Society Interface*, **6**(38), 735–747.

Wang, H., Zeng, L., Liu, H. & Yin, C. (2003). Measuring wing kinematics, flight trajectory and body attitude during forward flight and turning maneuvers in dragonflies. *Journal of Experimental Biology*. doi:10.1242/jeb.00183

Wang, S., Zhang, X., He, G. & Liu, T. (2014). Lift enhancement by dynamically changing wingspan in forward flapping flight. *Physics of Fluids*. doi:10.1063/1.4884130

Wang, Z. J. (2000). Vortex shedding and frequency selection in flapping flight. *Journal of Fluid Mechanics*. doi:10.1017/S0022112099008071

Wang, Z. J. (2005). Dissecting insect flight. *Annual Review of Fluid Mechanics*, **37**(1), 183–210.

Wang, Z. J., Birch, J. M. & Dickinson, M. H. (2004). Unsteady forces and flows in low Reynolds number hovering flight: Two-dimensional computations vs robotic wing experiments. *Journal of Experimental Biology*, **207**, 449–460.

Wang, Z. J., & Russell, D. (2007). Effect of forewing and hindwing interactions on aerodynamic forces and power in hovering dragonfly flight. *Physical Review Letters*. doi:10.1103/PhysRevLett.99.148101

Warrick, D. R., Tobalske, B. W. & Powers, D. R. (2005). Aerodynamics of the hovering hummingbird. *Nature*. doi:10.1038/nature03647

Weis-fogh, T. (1973). Quick estimates of flight fitness in hovering animals, including novel mechanism for lift production. *Journal of Experimental Biology*.

Windsor, S. P., Bomphrey, R. J. & Taylor, G. K. (2014). Vision-based flight control in the hawkmoth Hyles lineata. *Journal of the Royal Society Interface*, **11**(91), 20130921.

Witton, M. P., & Habib, M. B. (2010). On the size and flight diversity of giant pterosaurs, the use of birds as pterosaur analogues and comments on pterosaur flightlessness. *PLoS ONE*. doi:10.1371/journal.pone.0013982

Wodinsky, S. (2018, October). This robotic jellyfish could help save our reefs from climate change. NBC News. Retrieved from www.nbcnews.com/mach/science/robotic-jellyfish-could-help-save-our-reefs-climate-change-ncna913111.

Wood, R. J. (2007). Liftoff of a 60mg flapping-wing MAV. In *2007 IEEE/RSJ International Conference on Intelligent Robots and Systems*, Washington, DC: Institute of Electrical and Electronics Engineers, pp. 1889–1894.

Wootton, R. J. (1979). Function, homology and terminology in insect wings. *Systematic Entomology*, **4**(1), 81–93.

Wootton, R. J. (1981). Support and deformability in insect wings. *Journal of Zoology*. doi:10.1111/j.1469-7998.1981.tb01497.x

Wootton, R. J. (1992). Functional morphology of insect wings. *Annual Review of Entomology*. doi:10.1146/annurev.ento.37.1.113

Wu, J. H., & Sun, M. (2012). Floquet stability analysis of the longitudinal dynamics of two hovering model insects. *Journal of the Royal Society Interface*, **9**(74), 2033–2046.

Wu, J. H., Zhang, Y.-L. L. & Sun, M. (2009). Hovering of model insects: Simulation by coupling equations of motion with Navier-Stokes equations. *Journal of Experimental Biology*, **212**(20), 3313–3329.

Wu, P., Stanford, B. K., Sällström, E., Ukeiley, L. & Ifju, P. (2011). Structural dynamics and aerodynamics measurements of biologically inspired flexible flapping wings. *Bioinspiration & Biomimetics*, **6**(1), 016009.

Xie, C. M., & Huang, W. X. (2015). Vortex interactions between forewing and hindwing of dragonfly in hovering flight. *Theoretical and Applied Mechanics Letters*. doi:10.1016/j.taml.2015.01.007

Yates, G. T. (1986). Dragonfly aerodynamics. *Science*, **231**(4733), 10.

Yin, B., & Luo, H. (2010). Effect of wing inertia on hovering performance of flexible flapping wings. *Physics of Fluids*, **22**, 111902.

Young, J., Walker, S. M., Bomphrey, R. J., Taylor, G. K. & Thomas, A. L. R. (2009). Details of insect wing design and deformation enhance aerodynamic function and flight efficiency. *Science*, **325**(5947), 1549–1552.

Yu, Y., & Guan, Z. (2015). Learning from bat: Aerodynamics of actively morphing wing. *Theoretical and Applied Mechanics Letters*. doi:10.1016/j.taml.2015.01.009

Zhang, Y.-L., & Sun, M. (2010). Dynamic flight stability of hovering model insects: Theory versus simulation using equations of motion coupled with Navier-Stokes equations. *Acta Mechanica Sinica*, **26**(4), 509–520.

Zhao, L., Huang, Q., Deng, X. & Sane, S. P. (2010). Aerodynamic effects of flexibility in flapping wings. *Journal of the Royal Society*, 7(44), 485–497.

Zheng, Y., Wu, Y. & Tang, H. (2015). Force measurements of flexible tandem wings in hovering and forward flights. *Bioinspiration & Biomimetics*. doi:10.1088/1748-3190/10/1/016021

Zheng, Y., Wu, Y. & Tang, H. (2016a). A time-resolved PIV study on the force dynamics of flexible tandem wings in hovering flight. *Journal of Fluids and Structures*. doi:10.1016/j.jfluidstructs.2015.12.008

Zheng, Y., Wu, Y. & Tang, H. (2016b). An experimental study on the forewing-hindwing interactions in hovering and forward flights. *International Journal of Heat and Fluid Flow*. doi:10.1016/j.ijheatfluidflow.2015.12.006

Zhou, J., Adrian, R. J., Balachandar, S. & Kendall, T. M. (1999). Mechanisms for generating coherent packets of hairpin vortices in channel flow. *Journal of Fluid Mechanics*, 387, 353–396.

Zussman, E., Yarin, A. & Weihs, D. (2002). A micro-aerodynamic decelerator based on permeable surfaces of nanofiber mats. *Experiments in Fluids*. doi:10.1007/s00348-002-0435-6

Elements of Aerospace Engineering

Vigor Yang

Georgia Institute of Technology

Vigor Yang is the William R. T. Oakes Professor in the Daniel Guggenheim School of Aerospace Engineering at Georgia Tech. He is a member of the US National Academy of Engineering and a Fellow of the American Institute of Aeronautics and Astronautics (AIAA), American Society of Mechanical Engineers (ASME), Royal Aeronautical Society (RAeS), and Combustion Institute (CI). He is currently a co-editor of the Cambridge University Press Aerospace Series and co-editor of the book *Gas Turbine Emissions* (Cambridge University Press, 2013).

Wei Shyy

Hong Kong University of Science and Technology

Wei Shyy is President of Hong Kong University of Science and Technology and a Chair Professor of Mechanical and Aerospace Engineering. He is a fellow of the American Institute of Aeronautics and Astronautics (AIAA) and the American Society of Mechanical Engineers (ASME). He is currently a co-editor of the Cambridge University Press Aerospace Series, co-author of *Introduction to Flapping Wing Aerodynamics* (Cambridge University Press, 2013) and co-editor in chief of *Encyclopedia of Aerospace Engineering*, a major reference work published by Wiley-Blackwell.

About the Series

An innovative new series focusing on emerging and well-established research areas in aerospace engineering, including advanced aeromechanics, advanced structures and materials, aerospace autonomy, cyber-physical security, electric/hybrid aircraft, deep space exploration, green aerospace, hypersonics, space propulsion, and urban and regional air mobility. Elements will also cover interdisciplinary topics that will drive innovation and future product development, such as system software, and data science and artificial intelligence.

Cambridge Elements=

Elements of Aerospace Engineering

Elements in the Series

Distinct Aerodynamics of Insect-Scale Flight
Csaba Hefler, Chang-kwon Kang, Huihe Qiu and Wei Shyy

A full series listing is available at: www.cambridge.org/EASE

Printed in the United States
by Baker & Taylor Publisher Services